Dynamics and Biogeochemistry of River Corridors and Wetlands

Dynamics and Biogeochemistry of River Corridors and Wetlands

Edited by

LOUISE HEATHWAITE
Centre for Sustainable Water Management, University of Lancaster, UK

BRUCE WEBB
Department of Geography, University of Exeter, UK

DON ROSENBERRY
US Geological Survey, Denver, Colorado, USA

DAVID WEAVER
Department of Agriculture, Albany, Western Australia

MASAKI HAYASHI
Department of Geology and Geophysics, University of Calgary, Canada

The proceedings of the International Symposium on Dynamics and Biogeochemistry of River Corridors and Wetlands (S4) held during the Seventh Scientific Assembly of the International Association of Hydrological Sciences (IAHS) at Foz do Iguaço, Brazil, 3–9 April 2005. The symposium was organized by the International Commissions on Water Quality (ICWQ), Surface Water (ICSW) and Groundwater (ICGW).

IAHS Publication 294
in the IAHS Series of Proceedings and Reports

Published by the International Association of Hydrological Sciences 2005

IAHS Publication 294
ISBN 1-901502-03-1

British Library Cataloguing-in-Publication Data.
A catalogue record for this book is available from the British Library.

The papers included in this volume have been reviewed and some were extensively revised before publication.

IAHS is indebted to the employers of the Editors for the invaluable support and services provided that enabled them to carry out their task effectively and efficiently.

Publications in the series of Proceedings and Reports are available from:
IAHS Press, Centre for Ecology and Hydrology, Wallingford, Oxfordshire OX10 8BB, UK
tel.: +44 1491 692442; fax: +44 1491 692448; e-mail: jilly@iahs.demon.co.uk

Printed in The Netherlands by Krips BV, Meppel.

Preface

The river corridor is a diverse zone in terms of its scale, situation and character, and is one which is heavily modified by human activity. In many countries, river management has been driven by flood defence requirements, land drainage and also the desire of riparian landowners to have stable, well-defined river banks. As a result, channel cleaning, embanking, reinforcement and straightening have been widespread because they maximize the flood conveyance of the channel and maintain an essentially stable river course. As a consequence of these activities the special attributes of river corridors have in the past been largely ignored by river managers. Fortunately, recent attempts to improve "ecosystem health" in freshwater environments has meant that river conservation and rehabilitation measures have sought to re-introduce some of the attributes of river corridors and wetlands in terms of, for example, their sediment and solute attenuating capacity and their role as temporary flood storage areas.

River corridors span a wide scale. At one end of the spectrum, the corridor may comprise a narrow strip of riparian vegetation separating the hillslope from the river channel. At the other end, the corridor may consist of a wide and complex flood plain including many landscape units. The natural function of the corridor is to mediate the transport of water and dissolved and particulate material to the river network. The sedimentological and biogeochemical functions that the river corridor provide are particularly important to the long-term restoration and conservation of the river habitat. In research terms, there is a need to understand better the biogeochemical processes and transportation involved, including the manner in which the river corridor acts as a filter for particulate materials and as a sink for nutrients and other materials in soluble form, and how these fractions are affected by complex interactions between surface, hyporheic and groundwater flow in this environment. Better understanding of river corridor and wetland processes will help ensure that measures such as the introduction of riparian buffer strips (bands of well-developed riparian vegetation along the channel margin, including wetlands) by river managers, in the belief that they can help to process diffuse contaminant runoff from agricultural land, are more likely to succeed. Such buffer strips intercept eroded sediment and sediment-associated contaminants in overland flow and may remove soluble and colloidal nutrient forms from water draining laterally into the river channel. However, these processes are complex and thus the success of such buffer strips in attenuating and/or modifying nutrient transfer is still uncertain.

This edited volume addresses research issues focused on the biogeochemistry of the riverbank with special reference to data observation, modelling and attempts to restore these environments. The contents have been subdivided into a number of sections with specific themes. Thus there are a number of papers that focus on hydrochemistry, in particular, the role of organic carbon in upland wetlands and in flood plain rivers. A second section focuses on nutrients in the river corridor and wetland environment, where the emphasis is on understanding nitrogen and phosphorus dynamics. A large number of papers examine the river corridor zonation attributes such as seasonal water table fluctuation and vegetation dynamics. A final section of papers look at river corridors and ecosystem health.

The editors particularly wish to thank Dr Lucy Bolton for her assistance with putting together this volume.

Louise Heathwaite

Centre for Sustainable Water Management, The Lancaster Environment Centre
University of Lancaster, Lancaster LA1 4YQ, UK

Bruce Webb

Department of Geography, University of Exeter
Amory Building, Rennes Drive, Exeter EX4 4RJ, UK

Don Rosenberry

US Geological Survey, MS 413, Bldg 53, Box 25046, Denver Federal Center
Lakewood, Colorado 80225, USA

David Weaver

Department of Agriculture, 444 Albany Highway, Orana
Albany, Western Australia 6330, Australia

Masaki Hayashi

Department of Geology and Geophysics, University of Calgary
Calgary, Alberta T2N 1N4, Canada

Contents

1 Organic Carbon

Sulphate controls on dissolved organic carbon dynamics in blanket peat: linking field and laboratory evidence

PIPPA J. CHAPMAN[1], JOANNA M. CLARK[1], A. LOUISE HEATHWAITE[2], JOHN K. ADAMSON[3] & STUART N. LANE[4]

1 *School of Geography, University of Leeds, Leeds LS2 9JT, UK*
p.j.chapman@leeds.ac.uk

2 *Centre for Sustainable Water Management, The Lancaster Environment Centre, Lancaster University, Lancaster LA1 4YQ, UK*

3 *Centre for Ecology and Hydrology, The Lancaster Environment Centre, Lancaster LA1 4AP, UK*

4 *Department of Geography, University of Durham, Durham DH1 3LE, UK*

Abstract The dynamics of dissolved organic carbon (DOC) in blanket peat soil solution has been monitored at an upland catchment in northern Britain in relation to key environmental controls. Dissolved organic carbon (DOC) displayed a strong seasonal cycle that showed a clear relationship with soil temperature. However, in years of significant water table drawdown (>25 cm), summer DOC concentrations were substantially lower than in "normal" years, suggesting that factors other than temperature controlled the release of DOC. Major episodic release of sulphate (SO_4^{2-}), and an associated decline in pH, was also observed during drought periods. Controlled laboratory experiments were conducted over a 22 week period to factor out the relationships between water table depth, temperature and soil solution chemistry on DOC dynamics. The results from the laboratory experiments closely followed the patterns observed in the field data. Results suggest that changes in soil solution chemistry (especially SO_4^{2-}), in response to changes in soil hydrology, exerted a stronger control on DOC dynamics under drought conditions than variations in temperature alone.

Key words dissolved organic carbon; drought; peat; pH; sulphate; temperature; water table

INTRODUCTION

The sensitivity of DOC to changes in environmental conditions has been the topic of much debate lately, given recent observations of increased DOC concentrations in surface waters across Europe and North America (e.g. Skjelkvale *et al.*, 2001). Whilst some researchers have claimed that the rise in DOC is in response to increased biological activity associated with increasing air temperature (e.g. Freeman *et al.*, 2001), others have questioned this and pointed to the importance of other climatic factors such as changes in runoff regime (e.g. Pastor *et al.*, 2003) or frequency and magnitude of water table drawdown (e.g. Tipping *et al.*, 1999). However, great uncertainty still exists regarding which factors are important in controlling DOC dynamics in both the short and long-term.

Although it is widely acknowledged that there is a strong link between DOC production and temperature, the possibility of increases in global temperature driving the change in DOC has been questioned recently, given the magnitude of DOC rise; in the UK surface waters DOC has increased by an average of 91% during the last 15 years (1988–2002) (Evans *et al.*, 2004); this compares with a change in mean temperature of <1°C (Freeman *et al.*, 2004). In addition, research investigating the role of lower water tables on DOC production and export are contradictory, with some studies observing an increase (e.g. Tipping *et al.*, 1999) whilst others a decrease in DOC (e.g. Freeman *et al.*, 2004) or no significant change (e.g. Blodau *et al.*, 2004). Few have considered the impact of soil water chemistry on DOC mobility, even though marked changes in both pH and ionic strength have been observed during droughts in response to the oxidation of reduced sulphur to sulphate (SO_4^{2-}) (e.g. Adamson *et al.*, 2001), and the solubility of DOC is known to be influenced by pH and ionic strength (Thurman, 1985)

Whilst many studies have examined the effect of different environmental variables on DOC production, few have tried to separate out the effects of temperature and changes in water table in a quantitative way. A clear understanding of the nature of the relationships between DOC and the environmental factors controlling its production and release, is fundamental to gain a clearer understanding of the impact of environmental change on the export of DOC from peatlands. In this paper, the impact of peat hydrology and temperature on changes in soil solution chemistry and DOC dynamics is examined through integrated field and laboratory studies.

METHODS

Field site, monitoring and data collection

Fieldwork was conducted at Moor House-Upper Teesdale National Nature Reserve, which covers an area of 3500 ha in the North Pennine uplands of the UK (54°7′N, 2°4′W). This is a cool and wet upland site (290–848 m a.m.s.l), with a mean annual temperature (at 550 m a.m.s.l) of 5.3°C (1931–2000) and rainfall of 1982 mm year^{-1} (1934–1972) (Holden & Adamson, 2001). Approximately 80% of the reserve is covered with a deep (up to 4 m) raw oligo-fibrous peat, belonging to the Winter Hill Series. Vegetation cover is dominated by *Sphagnum* spp., *Eriophorum* spp. and *Calluna vulgaris* assemblages.

Routine monitoring has been conducted on the site since 1992 by the Environmental Change Network (ECN) in line with carefully prepared and published protocols (Sykes & Lane, 1996). Environmental variables were monitored at frequencies from 15 min (discharge) to 20 years (soil solid phase) (see Sykes & Lane, 1996), of which hourly water table and soil temperature readings in addition to biweekly soil solution data are presented here. Soil solution samples were collected from a network of six Pernart Super Quartz suction lysimeters at 10 cm depth at a suction of 50 kPa (Adamson *et al.*, 1998). Water samples were analysed for 14 constituents, of which pH, DOC and SO_4^{2-} are of interest to this paper. pH was determined using a low ionic strength probe, DOC by persulphate-ultraviolet oxidation (Skalar Auto Analyser), and SO_4^{2-} by ion chromatography (Dionex). Water table (one sensor) and soil temperature (three sensors at 1, 10 and 30 cm depth) were measured hourly. A map of sampling locations for the data utilized in this study can be found in Adamson *et al.* (1998).

Laboratory experiment

Twenty-four intact peat mesocosms (15-cm diameter, surface vegetation to 55 cm depth in peat) were collected from a location close to the soil solution sample site. These peat cores were taken to the laboratory and placed in a constant temperature room at 10°C (mean summer soil temperature at 10 cm depth), and subjected to a 12 h light–dark cycle. Rhizon soil moisture samplers were inserted at 1, 10, 20, 30, 40 and 50 cm depth within the peat mesocosm, with samples extracted for analysis at biweekly intervals at 100 kPa using a 10 ml syringe. Water samples were analysed in line with ECN protocols (Sykes & Lane, 1996) for pH, conductivity, SO_4^{2-} by ion chromatography (Dionex) and DOC by combustion-infrared analysis (Thermalox TOC/TN). Although practical restrictions due to equipment availability have caused slight variation in some methods between field and laboratory studies, care was taken to ensure compatibility by comparative analysis between laboratories (see Clark, 2005). Cores were left saturated (water table at 0 cm depth) for 3 months prior to the experimental manipulation to allow the soil system to equilibrate with the new laboratory conditions. Each core was connected to an external reservoir tube containing de-ionized water (4.5 cm in diameter, 65 cm length) by a series of plastic tubes, similar to the experimental design presented by Freeman *et al.* (1993). The water table depth in each core was adjusted by changing the elevation of the water table within the reservoir tube. Water tables were set at 0 (control), 15, and 45 cm depth for a period of 10 weeks and then rewetted to 0 cm and monitored for an additional 12 weeks. Each treatment was replicated four times. The range of water table depths and period of drawdown were set within the observed range of summer water tables at this site (Evans *et al.*, 1999).

RESULTS AND DISCUSSION

Controls on DOC dynamics under field conditions

Dissolved organic carbon production is generally considered to increase seasonally with temperature and aeration following summer water table drawdown, with export occurring during wetter conditions in the autumn (Mitchell & McDonald, 1992; Tipping *et al.*, 1999). Figure 1 shows the seasonal variation in DOC concentration in relation to the seasonality in these commonly identified environmental drivers. Soil temperature displays a strong seasonal cycle that peaks in August and shows little inter-annual variation (Fig. 1(a)). Water table also displays a general seasonality (Fig. 1(b)) with drawdown below 10 cm only occurring between April and November, although there is a high degree of inter-annual variability and within year variation. As observed in other studies (e.g. Adamson *et al.*, 2001; Scott *et al.*, 1998), DOC displays a strong seasonal cycle (Fig. 1(c)), peaking in late summer. However, considerably more variation in water table and DOC concentration is observed between years (peak annual DOC maximum values range between 25 and 50 mg l^{-1}) than is found in temperature alone (<1°C inter-annual variation in mean seasonal temperate). In particular, the years with lowest summer DOC concentrations (1995, 1994 and 1999)

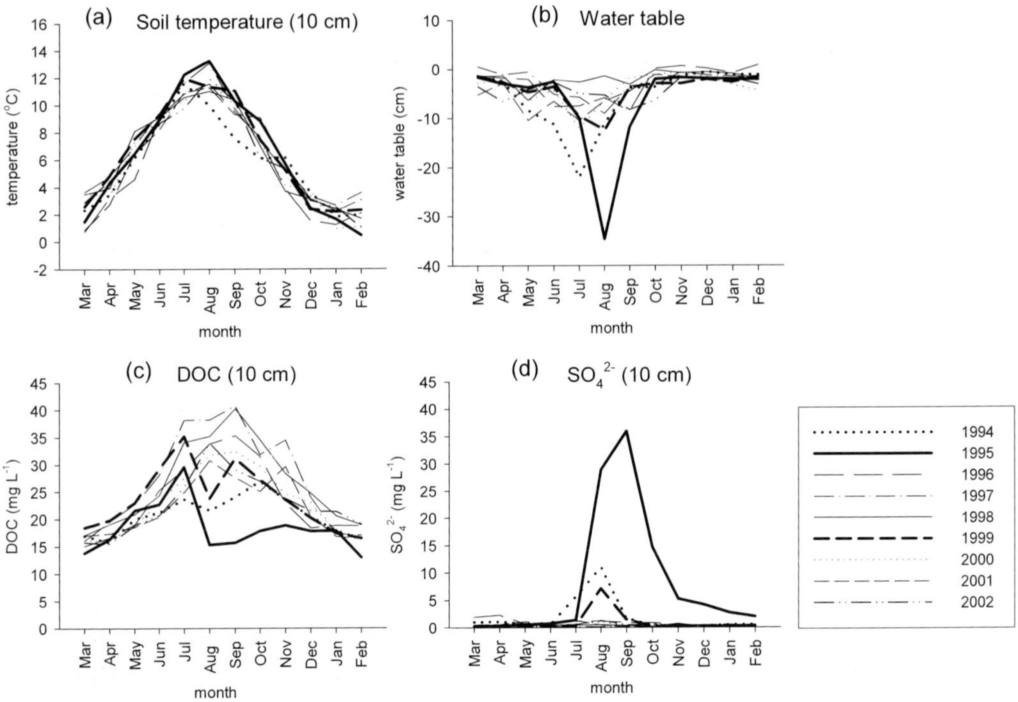

Fig. 1 Seasonal cycle in mean monthly (a) soil temperature at 10 cm depth, (b) water table, (c) dissolved organic carbon (DOC) concentration in soil solution at 10 cm depth, and (d) sulphate (SO_4^{2-}) concentration in soil solution at 10 cm depth.

correspond to the years of largest summer water table drawdown (1995, 1994 and 1999). These drought years also correspond to years with high sulphate (SO_4^{2-}) concentrations in soil solution (Fig. 1(d)).

Other studies have observed lower DOC concentrations in streams during drought years, which have been attributed to low flows (Pastor *et al.*, 2003) and difficulty in rewetting peat in response to physico-chemical changes in structure that cause a lag between production and export of 3–4 years (Watts *et al.*, 2001; Worrall *et al.*, 2003). However, these interpretations were based on stream data alone. Soil solution data presented here shows: (a) low concentrations of DOC during drought years in soil solution, and (b) a minimum in DOC concentrations in February/March following drought years. (Fig. 1(c)). Therefore, these observations in soil solution imply that low flows and difficulty in re-wetting the peat are unlikely to provide adequate explanation of the variations in DOC.

The field data also shows that extreme drying resulted in the release of SO_4^{2-} (Fig. 1(d)). Other studies have also observed an increase in SO_4^{2-} in soil and stream water following periods of drought (e.g. Scott *et al.*, 1998; Adamson *et al.*, 2001), which has been attributed to the mineralization and oxidation of organic sulphur and/or oxidation of reduced sulphur. Many of these studies have also noted a corresponding reduction in pH and increase in ionic strength (e.g. Adamson *et al.*, 2001), both of which hinder the dissociation of organic acids and so decrease their solubility (Thurman, 1985).

Although obvious seasonality exists in temperature, water table and DOC, it is difficult to determine whether changes in temperature and/or changes in water table are most crucial in controlling DOC dynamics. In addition, the observed decrease in DOC concentrations during drought years is contrary to the established theory, which states that increases in both temperature and aeration should lead to an increase in DOC production. Whilst many experiments have been conducted under changing temperatures (e.g. Freeman *et al.*, 2001) and simulated drought (e.g. Freeman *et al.*, 1993), little attention has been given to understanding the effect of specific water table depths on DOC production. Although the release of SO_4^{2-} and associated changes in pH and ionic strength have been observed (e.g. Adamson *et al.*, 2001), there has been no real consideration of the link between the dynamics of SO_4^{2-} and DOC under drought conditions. Therefore the controlled laboratory experiments described earlier (see Methods) were conducted to elucidate the relationship between changes in temperature and water table on soil solution chemistry (i.e. SO_4^{2-} and DOC), in order to improve understanding of the processes driving observed DOC dynamics in the field.

DOC dynamics under controlled laboratory conditions

Figure 2 shows how SO_4^{2-} and DOC concentrations in soil solution at 10 cm depth varied in response to different water table depths at a constant temperature. When the water table remained at the surface (0 cm) both SO_4^{2-} and DOC concentrations varied little over the 22 week period; SO_4^{2-} concentrations were low, generally less than 1 mg l^{-1}, whereas DOC concentrations displayed a small increase over time. In contrast, water table drawdown of 15 (mean summer drawdown) and 45 cm (maximum summer drawdown) lead to an increase in SO_4^{2-} and a decline in DOC. Rewetting the cores after 10 weeks lead to a rapid decline in SO_4^{2-} and increase in DOC. However, the decline in SO_4^{2-} was more rapid in the cores that had experienced a 15 cm water table drawdown, which resulted in a more rapid increase in DOC in these cores compared to the cores that had experienced a 45 cm water table drawdown. By week 22, SO_4^{2-} concentrations in all cores had returned to similar concentrations observed in week 0. In contrast, DOC concentrations in week 22 were over 10 mg l^{-1} higher than those observed in week 0 in the cores that had experienced water table drawdown. Figure 3 shows that the increase in SO_4^{2-} concentration in both the field and laboratory data resulted in an increase in H^+ ions, which are known to reduce the solubility of DOC (Thurman, 1985).

The results from this experiment show that a fall in water table and thus an increase in aeration lead to an increase in SO_4^{2-} concentrations, as was observed in the field data, due to the oxidation of reduced inorganic/organic sulphur. Although aeration increased DOC production, indicated by higher DOC concentrations after re-wetting (Fig. 2), this effect was suppressed during the drought by the release of SO_4^{2-}. Therefore, while temperature and aeration do control DOC production, changes in soil solution chemistry, related to sulphur dynamics, during drought years had the largest influence on DOC concentrations.

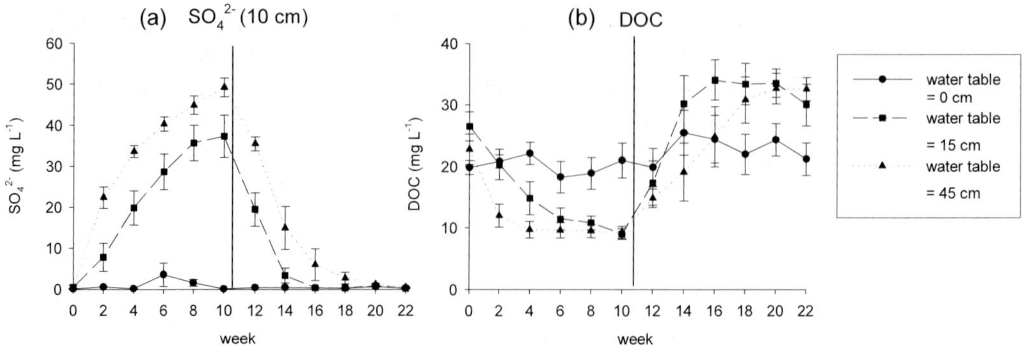

Fig. 2 Changes in (a) sulphate (SO_4^{2-}) concentrations and (b) dissolved organic carbon (DOC) concentrations in soil solution at 10 cm depth in response to changes in water table. Water table drawn-down from week 0 to 10, rewetted just after week 10 to week 22. Vertical line shows point of rewetting. Error bars show standard error where $n = 4$.

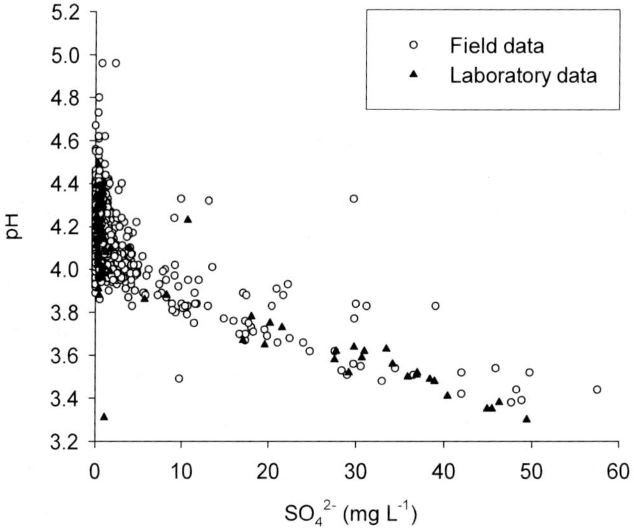

Fig. 3 The relationship between sulphate (SO_4^{2-}) concentration and pH in soil solution at 10 cm depth in the field and laboratory soil cores.

CONCLUSIONS

This paper shows that laboratory studies can be successfully used to factor out the relationships between temperature, water-table drawdown and DOC dynamics when carefully integrated within a long-term monitoring campaign in the field. Overall these combined studies show that peat hydrology had a stronger control on DOC dynamics during droughts through its influence on soil solution chemistry than changes in temperature alone. These results have implications for the future, as climate change scenarios predict an increase in the magnitude and frequency of drought events in northern Europe and America.

Acknowledgements This research has been supported by a NERC-CASE Studentship with CEH Lancaster (NER/S/A/2000/03431). The data used in this paper were supplied by the UK Environmental Change Network (www.ecn.ac.uk). We thank English Nature for allowing us to use the reserve for this work.

REFERENCES

Adamson, J. K., Scott, W. A. & Rowland, A. P. (1998) The dynamics of dissolved nitrogen in a blanket peat dominated catchment. *Environ. Pollut.* **99**, 69–77.

Adamson, J. K., Scott, W. A., Rowland, A. P. & Beard, G. R. (2001) Ionic concentration in a blanket peat bog in northern England and correlations with deposition and climate variables. *Eur. J. Soil Sci.* **52**, 69–79.

Blodau, C., Basiliko, N. & Moore, T. R. (2004) Carbon turnover in peatland mesocosms exposed to different water table levels. *Biogeochemistry* **67**, 331–351.

Clark, J. M. (2005) Environmental controls on the production and export of dissolved organic carbon from an upland peat catchment. PhD Thesis, School of Geography, University of Leeds, UK.

Evans, C. D., Monteith, D. T. & Copper, D. (2004) Long-term increase in dissolved organic carbon: observations, possible causes and environmental impacts. *Environ. Pollut.* (in press).

Evans, M. G., Burt, T. P., Holden, J. & Adamson, J. K. (1999) Runoff generation and water table fluctuations in blanket peat: evidence from UK data spanning the dry summer of 1995. *J. Hydrol.* **221**, 141–160.

Freeman, C., Hawkins, J., Lock, M. A. & Reynolds, B. (1993) A laboratory perfusion system for the study of biogeochemical responses of wetlands to climatic change. In: *Wetlands and Ecotones: Studies on Land–Water Interactions*, 75–83. National Institute of Ecology, New Delhi, India.

Freeman, C., Evans, C. D., Monteith, D, T., Reynolds, B. & Fenner, N. (2001) Export of organic carbon from peat soils. *Nature* **412**, 785.

Freeman, C., Fenner, N., Ostle, N. J., Kang, H., Dowrick, D. J., Reynolds, B., Lock, M. A., Sleep, D., Hughes, S. & Hughes, J. (2004) Export of dissolved organic carbon from peatlands under elevated carbon dioxide levels. *Nature* **430**, 195–198.

Holden, J. & Adamson, J. K. (2001) Gordon Manley: An upland meteorological pioneer. *J. Met.* **26**, 369–374.

Mitchell, G. & McDonald, A. T. (1992) Discolouration of water by peat following induced drought and rainfall simulation. *Water Res.* **26**, 321–326.

Pastor, J., Solin, J., Bridgham, S. D., Updegraff, K., Harth, K., Weishampel, P. & Dewey, B. (2003) Global warming and the export of dissolved organic carbon from boreal peatlands. *Oikois* **100**, 380–386.

Scott, M. J., Jones, M. N., Woof, C. & Tipping, E. (1998) Concentrations and fluxes of dissolved organic carbon in drainage water from an upland peat system. *Environ. Intern.* **24**, 537–546.

Skjelkvale, B. L., Stoddard, J. K. & Andersen, T. (2001) Trends in surface water acidification in Europe and North America (1989–1998). *Water Air Soil Pollut.* **130**, 787–792.

Sykes, J. M. & Lane, A. M. J. (1996) *The United Kingdom Environmental Change Network: Protocols for Standard Measurement at Terrestrial Sites*. HMSO, London, UK.

Thurman, E. M. (1985) *Organic Geochemistry of Natural Waters*. Kluwer Academic Publishers Group, Dordrecht, The Netherlands.

Tipping, E., Woof, C., Rigg, E., Harrison, A. F., Ineson, P., Taylor, K., Benham, D., Poskitt, J., Rowland, A. P., Bol, R. & Harkness, D. D. (1999) Climatic influences on the leaching of dissolved organic matter from upland UK moorland soils, investigated by a field manipulation experiment. *Environ. Intern.* **25**, 83–95.

Watts, C. D., Naden, P. S., Machell, J. & Banks, J. (2001) Long term variation in water colour from Yorkshire catchments. *Sci. Total Environ.* **278**, 57–72.

Worrall, F., Burt, T. & Shedden, R. (2003) Long term records of riverine dissolved organic matter. *Biogeochemistry* **64**, 165–178.

Dynamics and Biogeochemistry of River Corridors and Wetlands (Proceedings of symposium S4 held during the Seventh IAHS Scientific Assembly at Foz do Iguaçu, Brazil, April 2005). IAHS Publ. 294, 2005.

Potential carbon losses from boreal pond and riparian areas: influence of temperature and drought

RICHARD M. PETRONE[1], KEVIN J. DEVITO[2], STEPHEN KAUFMAN[1], MERRIN L. MACRAE[2] & JAMES M. WADDINGTON[3]

1 *Department of Geography & Environmental Studies, and Cold Regions Research Centre, Wilfrid Laurier University, Waterloo, Ontario, Canada*
rpetrone@wlu.ca

2 *Department of Biological Sciences, University of Alberta, Edmonton, Alberta, Canada*

3 *School of Geography and Geology, McMaster University, Hamilton, Ontario, Canada*

Abstract Wetland-pond systems on the Western Boreal Plain (WBP) in northwestern Canada account for as much as 50% of the landscape and represent the largest carbon pool in Canada. However, this region is also among the most threatened due to increased industrial, agricultural and recreational development. As ponds dry due to enhanced evaporation and hydrological disconnection with upland areas, pond sediments are exposed and riparian areas dry out, leading to increased decomposition and the loss of CO_2 to the atmosphere. This paper demonstrates how respiration/oxidation potentials vary in this region, and are linked to the different organic substrates that are found in areas with variable surficial geology and vegetation cover. We illustrate carbon exchange in relation to vegetation succession and increased oxidation in the surrounding riparian peatlands in drying pond-wetland systems in the sub-humid climates of the boreal plain.

Key words boreal forest; carbon dioxide; northwest Canada; oxidation

INTRODUCTION

Boreal wetlands have accumulated significant amounts of carbon, largely in peat deposits, which may be threatened by increasing climate variability and escalated human development (Gorham, 1994; Waddington *et al.*, 2001). When compared to humid boreal regions, the continental Western Boreal Plain (WBP) of Canada has much more complex surface and groundwater hydrology due to deep and heterogeneous glacial deposits (Ferone & Devito, 2004). Further, the sub-humid climate of WBP makes this region more susceptible to climatic variations (Devito *et al.*, 2005). Currently the WBP is experiencing increased resource extraction (roads, seismic cutlines, forestry) and possible changes in climate that may alter the existence and functioning of the many wetland-pond complexes. It is therefore necessary to understand how factors that influence carbon accumulation and emissions will vary in the heterogeneous landscape of the WBP to better predict how these complexes will function while under climatic or developmental stress.

Previous studies have shown that wetland carbon fluxes will vary both temporally and spatially in response to a combination of physical factors (Bridgham *et al.*, 1999;

Bubier *et al.*, 2003), but largely those that influence peat moisture and temperature (Griffis *et al.*, 2000; Bubier *et al.*, 2003; Petrone *et al.*, 2003). Shifts in climate or land-use change that alter wetland hydrology exposing normally flooded peat to oxygen provide a potential for rapid organic matter breakdown and carbon emissions (Hogg *et al.*, 1992; van Seters & Price, 2001). However, some research has linked higher seasonal temperatures with an increase in plant productivity and therefore an increase in carbon sequestration (Waddington *et al.*, 1996).

Incubation experiments have commonly been used to determine CO_2 fluxes under oxic and anoxic conditions (e.g. Blodau & Moore, 2003). While applying differing temperatures is common (e.g. 4 and 20°C), applying varying moisture regimes is less so (e.g. Blodau & Moore, 2003). Utilizing both is rarely used (e.g. Waddington *et al.*, 2001), but further research is required considering the emerging evidence that no single variable can best predict changes in CO_2 fluxes.

The goal of this paper is to investigate the influence of temperature and moisture regime changes on CO_2 fluxes from riparian soils and pond sediments, which are the soil types most readily susceptible to drought conditions, in two WBP wetland-pond complexes with contrasting substrate types: one with little organic accumulation in riparian areas, and one with a well developed riparian peatland and pond with organic rich sediments. The laboratory incubations are used to infer how increased climate variability leading to longer drought periods may influence the respiration potential and loss of carbon from wetland-pond systems typical of the WBP. This objective will also address the feedback processes that may be responsible for the presence or absence of peat in some of the basins in this region. All of these factors are examined in sediments collected from wetland-ponds located in a topographic high region in the WBP where the water balance and levels will be the most susceptible to climate change (Ferone & Devito, 2004). This study is part of a larger project to ultimately understand the climatic and landscape influences on carbon sequestration and peatland-pond sustainability.

STUDY SITE

The studied wetlands are located in the Western Boreal Plains (WBP) region of Canada, near Utikuma Lake (57°N, 115°W), 300 km north of Edmonton, Alberta (Fig. 1). The physiography, climate and vegetation are described in detail in Ferone & Devito (2004), and Bayley & Prather (2003). The study area contains moraine, outwash, and glaciolacustrine deposits to depths of 20 to 100 m, underlain by Smoky Group shale (Vogwill, 1978). Peat-forming wetlands cover about 30% of the land surface (Vitt *et al.*, 2000). This low relief and poorly integrated drainage network results in a mosaic of shallow depression lakes, and ponds surrounded by peatland development, herein referred to as pond-peatland complexes.

Climate in the WBP is between the Mid- and High-Continental Boreal (NWWG, 1988) with average July and January temperatures of 16°C and –15°C, respectively (Environment Canada, 2004). Total precipitation is approximately 515 mm year^{-1}, with 25% as snowfall. In most years evaporation exceeds precipitation and as a result runoff is low (Devito *et al.*, 2005).

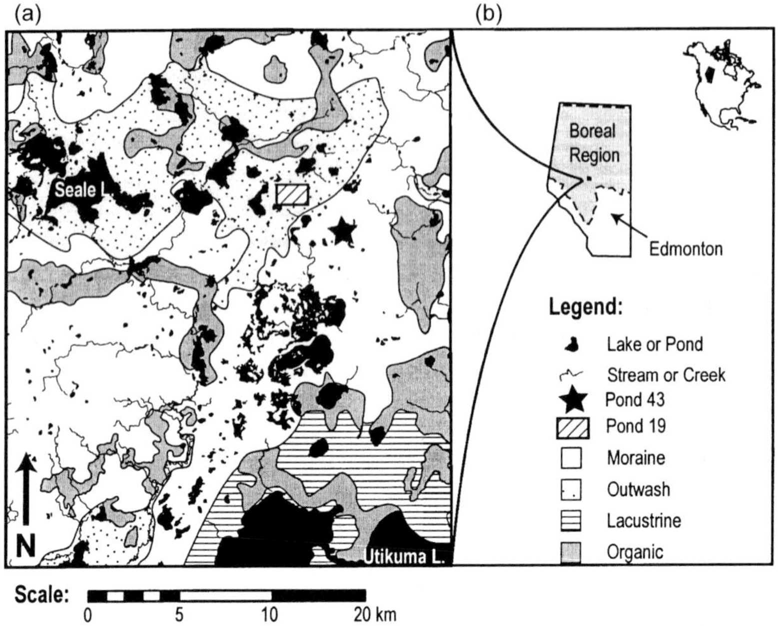

Fig. 1 (a) The Western Boreal Plains located within the Canadian interior plains with the Utikuma study area, Alberta. (b) The moraine and outwash plain study wetland-pond complexes located among the many boreal ponds and wetlands.

The two pond-peatland complexes under study are representative of much of the WBP, and were selected to represent topographic high regions where precipitation and evaporation dominate the water balance and the pond-peatland complexes function as groundwater recharge and are therefore most susceptible to changes in climate (Ferone & Devito, 2004).

Pond 43 is on an ice melt moraine with clay-silt rich till substrates. Pond water depth ranged from 0 to 100 cm over the past five years. Underlying the pond is 200 to 300 cm of long-term accumulation of organic rich pond sediments (gyttja) underlain by low permeability silt and clay substrates. The peatlands adjacent to the pond are 20 to 100 m wide and 150 to 300 cm in depth. The top 30 cm of peat is fibric, mesic to approximately 150 cm depth and mesic-humified to 300 cm (Ferone & Devito, 2004). *Sphagnum* spp., *Ledum groenlandicum*, and *Chamaedaphne calyculata* dominate the understory, while overstory species are *Picea mariana*, *Larix laricina* and some *Salix* spp. With recent drought and low water levels *Carex* spp., *Typha* spp., and *Salix* spp have extended inwards from the peat edge over exposed sediments.

Pond 19 is located in courser grained fluvial sediments, with pond water perched approximately 20 m above the regional groundwater. Pond water depths fluctuate between 50 and 150 cm, with mineral rich silty sediments accumulating over the top of 1–2 m of clay rich substrates, above sands and gravels. The riparian area adjacent to the pond ranges from 60 to 120 m, with organic accumulation of less than 20 cm underlain with 20–30 cm organic loam and clay-silt. *Typha* spp. are located in flooded sediments, *Carex* spp. dominate near the pond edge and grasses with *Salix* spp. and raspberry (*Rubus idaeus*) in higher elevations near the riparian–forest edge.

METHODS

Ten peat cores (~35 cm long) were collected in August 2003 along a pond sediment (flooded, exposed and re-vegetated), riparian, hillslope gradient from both wetland-pond complexes (19 and 43). At each site, duplicate cylindrical cores (7.62 cm ID) were taken manually at five locations using aluminium core tubes. All cores were sealed and frozen for transport, after which each core was divided into 5-cm sections for analysis, giving a total of 70 five-centimetre cylindrical sections. In addition, approximately 2 litres of pond and peat water was removed from each study pond for use in the incubations.

Core sections representative of the peat/soil profile were used for bulk density, specific yield, organic content and incubation experiments with samples taken from the surface (5–10 cm), mid-depth (15–20 cm) and deep (30–35 cm) peat/soil. Two temperature regimes (4°C and 20°C) and three moisture regimes (dry, moist and wet) incubations were chosen that represent the range of possible climate conditions that may occur in the study region. Each of the separated core sections were divided into three ~35 g samples and placed in 250 ml sample jars equipped with a septum lid. For the "saturated" samples, approximately 100 ml of pond water was added to each sample, while the "moist" samples received enough water to soak the sample without creating a slurry, and the "dry" samples were left at 80°C for 24 hours prior to the incubation. This technique resulted in a variable headspace in the jar depending on the initial bulk density of the peat sample. After the incubations were started, a syringe was used to collect 10 ml of headspace gas at an initial time, and gas collection was repeated every 12 hours for 48 hours. A total of 1800 syringes were collected and analysed within 60 hours on a Varian Star 5200 Gas Chromatograph.

RESULTS

Soil characteristics and substrate quality

Soil physical properties reflect the dominance of the mainly mineral riparian and pond soils at pond 19 and the presence of peat and organic deposits at pond 43. Soil and sediment organic content and porosity was higher, and specific yield and bulk density were lower at the moraine pond 43 compared to the perched pond 19 (Table 1). Overall, C:N ratios increase from pond sediments to riparian areas and are generally greater at pond 43 (Table 1). C:N ratios were less than 20 in both sites, with the exception of the riparian peat in pond 43.

Pond and riparian zone carbon oxidation rates

Under warm, moist conditions depth integrated CO_2 fluxes from flooded and exposed vegetated sediments, and riparian zone soils in pond 43 were greater than those from pond 19 (Fig. 2). At both sites CO_2 fluxes from bare sediments and vegetated sediments were higher than from the surrounding riparian soils or peat.

Table 1 Summary of pond sediment and riparian soil characteristics at three sample depths for the two pond-peatland complexes. Carbon:nitrogen (C:N) ratios, surface CO_2 flux and volumetric soil moisture are also shown for each site and transect location with bulk values representing the substrate surface.

Depth (cm)	Sediments: Pond 43	Pond 19	Vegetated sediments: Pond 43	Pond 19	Riparian: Pond 43	Pond 19
	Loss on ignition (%)					
5 – 10	39	16	57	13	79	57
15 – 20	25	12	72	10	91	33
30 – 35	48	10	65	8	87	4
	Specific yield					
5 – 10	0.06	0.19	0.05	0.05	0.09	0.08
15 – 20	0.06	0.15	0.06	0.05	0.09	0.06
30 – 35	0.08	0.06	0.09	0.05	0.09	0.07
	Porosity					
5 – 10	0.87	0.75	0.88	0.60	0.90	0.76
15 – 20	0.87	0.72	0.88	0.63	0.92	0.65
30 – 35	0.91	0.65	0.89	0.63	0.86	0.41
	Bulk density					
5 – 10	0.11	0.24	0.11	0.41	0.05	0.21
15 – 20	0.10	0.27	0.10	0.38	0.06	0.36
30 – 35	0.07	0.37	0.09	0.36	0.11	0.82
Bulk						
C:N	12	11	18	13	26	16
CO_2 (mg m^{-2} day^{-1})	6.8	4.7	25.0	19.7	52.5	32.2
Volumetric soil moisture	0.40	0.47	0.41	0.40	0.15	0.28

To assess the influence and interaction of temperature and moisture, vegetated and exposed sediment sites were merged and the two riparian zone sites were merged. Higher CO_2 emissions were observed in higher temperature incubations in almost all trials (Fig. 3). The influence of moisture exhibited a classic quadratic relationship, increasing from dry to moist conditions, with further reductions in CO_2 emissions in the saturated scenario. The interaction of moisture and temperature was such that the increase in CO_2 with temperature was greatest when the substrates were moist (Fig. 3).

For both ponds 43 and 19, sediments lost more CO_2 on average than riparian soils and showed more pronounced losses during warmer incubations. The exception to this trend is during saturated and cold conditions when pond 43 riparian and sediment areas sequestered CO_2. Pond 19 sediments also sequestered CO_2 under saturated conditions but in contrast to pond 43, riparian areas at pond 19 released CO_2 when saturated. Measured CO_2 fluxes during dry and moist conditions were greater in the riparian peat of pond 43 than the sedge/grass riparian zone of pond 19. This trend reversed for saturated conditions when pond 43 became a sink while pond 19 switched to a source (Fig. 3).

In situ data is shown (Table 1) from chamber measurements conducted on a warm, dry summer day in August 2002 (Petrone, unpublished data; cf. Petrone *et al.*, 2003). Table 1 shows that the CO_2 flux during a warm period is related to soil moisture, and decreases in CO_2 are observed from the sediments to the riparian zone. The trend with

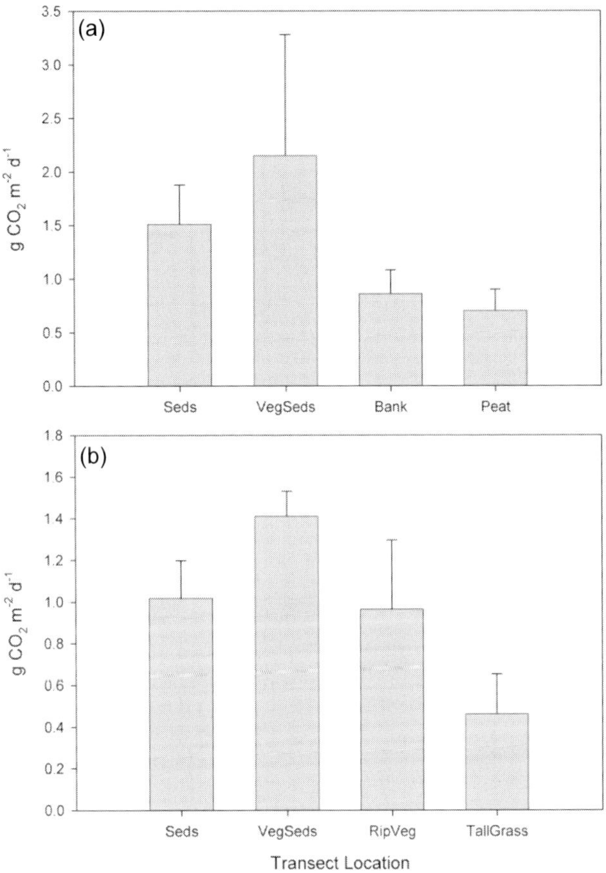

Fig. 2 Depth integrated mean CO_2 loss to the surface for each of the four transect locations at: (a) the Moraine Pond 43, and (b) the Outwash Pond 19. Error bars indicate standard deviation. "Seds" refers to the pond sediments; "VegSeds" to exposed pond sediments that have experienced some vegetation succession; "RipVeg", "TallGrass" and "Peat" all refer to substrate samples taken from the riparian zone; and "Bank" refers to lateral sampling into the peat bank along the pond perimeter.

soil moisture observed in the laboratory is also observed in the field, with an increase in CO_2 flux between saturated and moist soils. Furthermore, riparian *in situ* fluxes are much greater than with the pond sediments, but higher CO_2 fluxes were observed in vegetated than bare exposed pond sediments.

DISCUSSION

Of key interest in this study was whether the riparian soils or pond sediments behave differently within and between sites with contrasting organic accumulation. Aeration of exposed sediments at both ponds results in a greater loss of carbon than the riparian zone substrates at both sites. This indicates that pond sediments can potentially be a

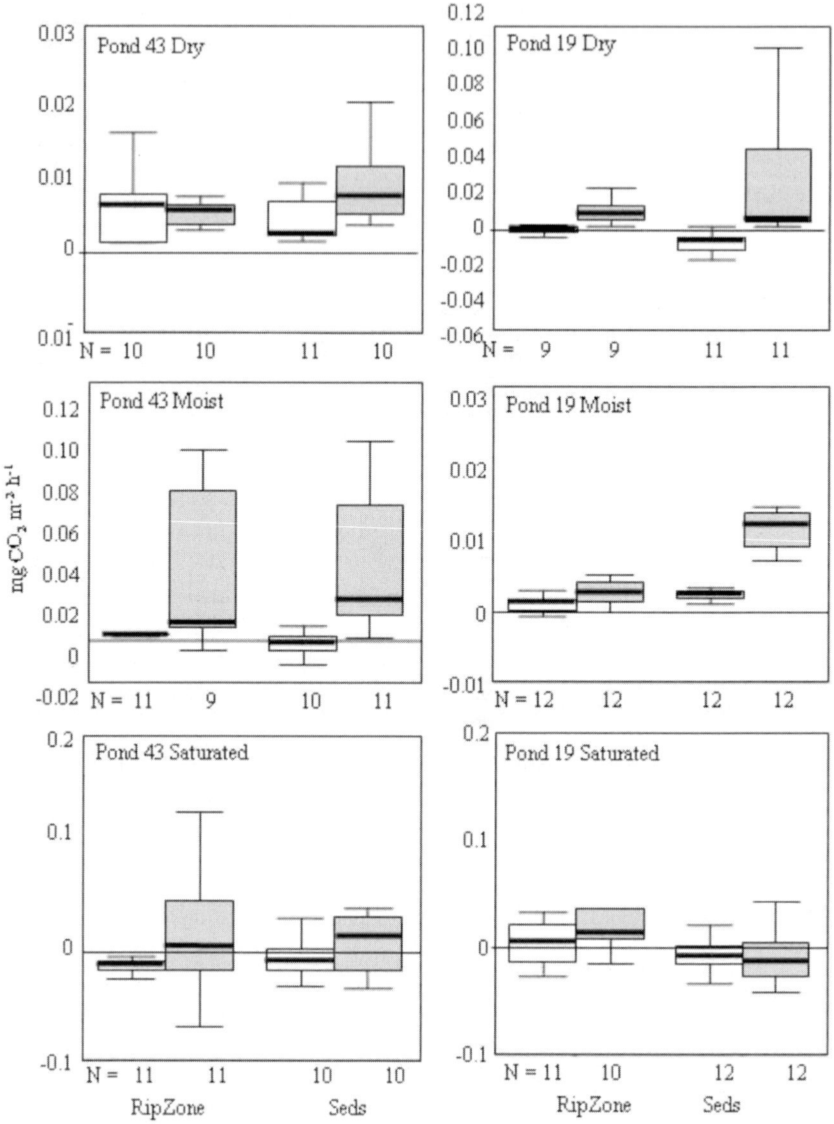

Fig. 3 Grouped depth-integrated oxidation rates for ponds 43 and 19 under dry, moist, and saturated conditions. White bars indicate the cooler temperature (4°C) and grey bars represent the warmer temperature (20°C) treatments. Boxes show the 25, 50 and 75th percentiles. The 10th and 90th percentiles are shown with whiskers. "RipZone" refers to all of the riparian zone samples grouped, and "Seds" refers to the grouping of all sediment samples (exposed included).

large source of carbon during prolonged drought. However, oxidation of pond sediments will result in subsidence and potentially re-flooding. As with eastern boreal pond-peatland complexes (Dillon & Molot, 1997) pond infilling may be sustained by prolonged flooding and anoxia.

There are potentially different responses to climatic variability by the two ponds with contrasting organic content. Both the pond sediments and riparian soils at pond 43 have similar CO_2 dynamics and seem to be exhibiting a large change when sediment conditions are manipulated (i.e. moisture and temperature). In contrast, the mineral-rich pond sediments appear to have a more pronounced response to environmental change than the riparian soils at pond 19, as they more readily shift from a source to sink. This is illustrated by a temperature change invoking a change in flux direction in the pond sediments under dry conditions (Fig. 3).

The two ponds respond to drought in a different manner. Under dry, cool conditions, sediments at pond 43 are a CO_2 source. However, as moisture increases and plants succeed into the pond, the CO_2 losses decrease in magnitude. In contrast, sediments at pond 19 are a CO_2 sink under cool, dry conditions, but become sources if conditions become more moist.

The contrasting CO_2 fluxes between the two ponds suggest that losses from these sites will be a function of substrate quality and the quantity of organic material. The riparian zone sites at pond 43 (peat dominated) are large sources of CO_2 during dry and warm conditions, but become sinks if conditions become very wet or saturated. This may reflect the poorer substrate quality (large C:N ratio). Basin morphometry and impermeable substrates that promote saturated conditions may explain long-term accumulation of peat at this site. In contrast, although CO_2 fluxes varied with temperature and moisture, there is an increase in carbon losses at pond 19 under warmer temperatures. Losses occurred in the riparian zone soils at pond 19 in all moisture and temperature conditions and may explain the lack of organic matter accumulation over time. In the riparian zone at pond 19, the tall grasses become strong sources of CO_2 if conditions are cool and wet suggesting the more significant role of the roots with this type of vegetation.

CONCLUSIONS

Spatial variability between riparian zone soils and pond sediments is less than the observed variability induced by temperature and moisture variations in organic-rich and mineral-rich wetland pond complexes in the Western Boreal Plain under laboratory experimental conditions. Oxidation rates are observed to be generally higher in organic riparian zones, but the contrast occurs when saturated conditions exist. Oxidation during cold saturated conditions may help to explain the lack of long-term accumulations of carbon at mineral dominated sites like pond 19.

An understanding of the hydrogeological controls on the amplitude and periodicity of water table fluctuations as it influences substrate moisture status is essential in these systems. Such controls and fluctuations can result in spatial variability of carbon accumulation rates as well as oxidation rates. Perched systems like pond 19 can have extensive lateral recharge which may limit the extent and periodicity of flooding in its riparian area. Thus, moisture conditions will be maintained in the moist to dry range. Conversely, the basin morphometry and lower permeability of the pond sediments at pond 43 will tend to maintain saturated conditions in the surrounding riparian peatland for longer periods. These prolonged saturated conditions may enhance or facilitate peat accumulation.

Finally, the interactions and influence of moisture and temperature status on oxidation rates demonstrate that a dry and warm climate will be very different from a wet and warm climate. Thus, along with the hydrogeology of the region we need to know the feedbacks associated with the storage changes in peat to truly understand the potential response of these systems to climate change.

Acknowledgements The authors wish to acknowledge the Natural Science and Engineering Research Council (NSERC) of Canada, Ducks Unlimited, Sustainable Forest Management Network and Alberta Pacific Forest Industries for their financial support on this project. The authors also thank J. R. VanHaarlem and Sarah Day for their assistance in the field and laboratory, and the anonymous reviewer for their insightful comments on this manuscript.

REFERENCES

Bayley, S. E. & Prather, C. M. (2003) Do wetland lakes exhibit alternative stable states? Submersed aquatic vegetation and chlorophyll in western boreal shallow lakes. *Limnology and Oceanography* **48** (6), 2335–2345.

Blodau, C. & Moore, T. R. (2003) Experimental response of peatland carbon dynamics to a water table fluctuation. *Aquatic Sciences* **65**, 47–62.

Bridgham, S. D., Pastor, J., Updegraff, K., Malterer, T. J., Johnson, K., Harth, C. & Chen, J. Q. (1999) Ecosystem control over temperature and energy flux in northern peatlands. *Ecological Applications* **9**(4), 1345–1358.

Bubier, J. L., Bhatia, G., Moore, T. R., Roulet, N. T. & Lafleur, P. M. (2003) Spatial and temporal variability in growing-season net ecosystem carbon dioxide exchange at a large peatland in Ontario, Canada. *Ecosystems* **6**(4), 353–367.

Devito, K. J., Creed, I. F. & Fraser, C. (2005) Controls on runoff from a partially harvested aspen forested headwater catchment, Boreal Plain, Canada. *Hydrol. Processes* (in press, accepted March 2004).

Dillon, P. J. & Molot, L. A. (1997) Dissolved organic and inorganic carbon mass balances in Central Ontario lakes. *Biogeochemistry* **36**, 29–42.

Environment Canada (2004) *Canadian Climate Normals or Averages 1971–2000.* http://www.climate.weatheroffice.ec.gc.ca/climate_normals/index_e.html

Ferone, J. M. & Devito, K. J. (2004) Shallow groundwater-surface water interactions in pond-peatland complexes along a Boreal Plains topographic gradient. *J. Hydrol.* **292**, 75–95.

Gorham, E. (1994) The future of research in Canadian peatlands: a brief survey with particular reference to global change. *Wetlands* **14**(3), 206–215.

Griffis, T. J., Rouse, W. R. & Waddington, J. M. (2000) Interannual variability of net ecosystem CO_2 exchange at a subarctic fen. *Global Biogeochemical Cycles* **14**(4), 1109–1121.

Hogg, E. H., Lieffers, V. J. & Wein, R. W. (1992) Potential carbon losses from peat profiles—effects of temperature, drought cycles and fire. *Ecological Applications* **2**(3), 298–306.

National Wetlands Working Group (NWWG) (1988) *Wetlands of Canada*. Ecological Land Classification Series no. 24, Environment Canada and Polyscience Publications, Ottawa, Canada.

Petrone, R. M., Waddington, J. M. & Price, J. S. (2003) The effects of restoration on ecosystem scale flux of CO_2 from a vacuum harvested peatland. *Wetlands Ecology and Management* **11**, 419–432.

van Seters, T. E. & Price, J. S. (2001) The impact of peat harvesting and natural regeneration on the water balance of an abandoned cutover bog, Quebec. *Hydrol. Processes* **15**, 233–248.

Vitt, D. H., Halsey, L. A., Bauer, I. E. & Campbell, C. (2000) Spatial and temporal trends in carbon storage of peatlands of continental western Canada through the Holocene. *Can. J. Earth Sci.* **37**(5), 683–693.

Vogwill, R. I. J. (1978) Hydrogeology of the Lesser Slave Lake area, Alberta. *Alberta Research Council Report 77.* Edmonton, Alberta, Canada.

Waddington, J. M. & Roulet, N. T. (1996) Atmosphere-wetland carbon exchanges: scale dependency of CO_2 and CH_4 exchange on the developmental topography of a peatland. *Global Biogeochemical Cycles* **10**(2), 233–245.

Waddington, J. M., Rotenberg, P. A. & Warren, F. J. (2001) Peat CO_2 production in a natural and cutover peatland: implications for restoration. *Biogeochemistry* **54**, 115–130.

Dynamics of organic carbon between the Amazon River and flood plain lakes

P. MOREIRA-TURCQ[1], G. BARROUX[2], M. BERNARDES[3], M.-P. BONET[2], L. MAURICE-BOURGOIN[2], M. PEREZ[3] & P. SEYLER[2]

1 *IRD–Institut de Recherche pour le Developpement, UMR LMTG–Departamento de Geoquímica, Universidade Federal Fluminense, Morro do Valonguinho s/nº, 24020-007 Niterói, Rio de Janeiro, Brazil*
pturcq@geoq.uff.br

2 *IRD–UMR LMTG, Université Paul Sabatier, 39 allées Jules Guesde, F-31000 Toulouse, France*

3 *Departamento de Geoquímica, Universidade Federal Fluminense, Morro do Valonguinho s/nº, 24020-007 Niterói, Rio de Janeiro, Brazil*

Abstract Water is constantly exchanged between river channels and flood plain lakes. Little is known about the flux at which organic carbon is transferred to and from lakes, and the residence time of flood plain storage. Flux has a special importance for the carbon cycle, especially in the Amazon basin. During four hydrological years organic carbon behaviour and its fluxes were studied in an Amazon flood plain and in the Amazon River. In the Amazon River, particulate organic carbon contents is very constant and represents approx. 2% of the total suspended matter. However, in Amazon flood plain lakes organic carbon content is very variable and represents 2–30% of the total suspended matter and can reach 40–50% of suspended matter during extensive phytoplankton blooms. Dissolved organic carbon is also very variable (5–30 mg l^{-1}) in flood plain lakes, while in the Amazon River its concentration is between 3 and 5 mg l^{-1}. Flood plain lakes are an important source of organic carbon to the Amazon River and are also a sink for Amazon particulate organic carbon.

Key words Amazon River; dynamics; flood plains; organic carbon; organic matter

INTRODUCTION

Flood plains or "várzeas" (Amazonian flood plains) are dynamic and complex wetland systems which periodically oscillate between terrestrial and aquatic phases. The most comprehensive approach to deal with flood plains is the flood pulse concept (Junk *et al.*, 1989) which is based on hydrological considerations of the river, its catchment area, and its flood plain. Tropical várzeas have a special interest, mainly because of their huge surface area, which can be useful land reserves for agriculture, animal husbandry and forestry, as well as areas for settling populations (Junk, 1997). They represent an important site for production, storage and export of carbon.

Water and sediments are constantly exchanged between river channels and flood plain lakes. Flood plains of large rivers are built by the formation of bars and the accumulation of diffuse sediment carried in over bank flows and in channelized flows (Dunne *et al.*, 1998). The rates at which sediment is transferred to and from flood plains and the residence time of várzea storage, are not well known. According to the flood pulse concept (Junk, 1997), most organic carbon processed in river flood plains is produced inside the flood plains. Inflow of the turbid, nutrient-rich water of the Amazon River onto flood plains helps sustain the flood plain's high fertility (Melack & Forsberg, 2001). In Amazon flood plains, main primary producers are trees, terrestrial and aquatic herbaceous plants, phytoplankton, and periphyton (Junk, 1997). The majority of the plankton (Richey, 1982) and the bulk of carbon (Junk, 1997) are probably degraded or are buried *in situ*. Flood plains are the largest source of atmospheric methane (Quay *et al*., 1992), where the estimated global source strength is approx. 110 Tg CH_4 $year^{-1}$ and the tropical flood plain contribution is approx. 60% (Bartlet & Harris, 1993). During floods, the Amazon flood plains are also an important natural source of CO_2 (Richey *et al*., 2002) for the atmosphere. Várzeas are considered as intermediate systems between open transporting systems (rivers) and closed (lakes) accumulative systems. The accumulation of material can occurs over short time periods or at intervals of hundreds (Moreira-Turcq *et al*., 2004) or thousands of years (Irion *et al*., 1999).

The mass balance of sediments (Dunne *et al*., 1998; Filizola, 1999) and carbon (Richey *et al*., 1990; Moreira-Turcq *et al*., 2003a) in the Amazon system show a loss (Ertel *et al*., 1986; Moreira-Turcq *et al*., 2003b) indicating the presence of a sink, mainly in várzea systems. On the other hand, organic carbon produced in the várzea lakes is very labile and is constantly exported to the Amazon River. The aim of our study is to evaluate the dynamic and the main biogeochemical processes involving organic carbon in lower Amazon flood plain lakes.

STUDY AREA

The Amazon basin drains a surface area of around 6 000 000 km^2, constituting approx. 5% of the Earth's terrestrial surface. The Amazon River and its large tributaries are accompanied along their middle and lower courses by large flood plains, which cover an area of about 300 000 km^2 (Junk, 1997), including 110 000 km^2 in the main stream (Melack & Fisher, 1990). Permanent and temporary lakes increase in size and become connected to each other during high river discharge. The flood plain "Várzea do Lago Grande de Curuai" (Fig. 1) is located on the right bank of the Amazon River, across from the city of Óbidos, 900 km from the mouth. It is a complex system of more than 30 interconnected lakes, with a maximum inundated area of 2300 km^2, representing 1% of the total flooded area of the Central Amazon basin. The catchments of the Várzea do Lago Grande de Curuai cover an area of 3660 km^2, including open water areas. The Várzea do Lago Grande de Curuai is comprised of several (white-water lakes) waters characterized by high suspended sediment loads and (black-water lakes) waters characterized by high concentrations of dissolved humic acids and low concentrations of suspended sediment interconnected with each other and permanently connected to the Amazon mainstream by different small channels. The major lake, Lago Grande de Curuai, is a white-water lake of approx. 359 km^2, located between 56°00′W (upstream) and 55°03′W (downstream) and 2°17′S and 1°55′S. This lake is permanently linked to the Amazon River by two channels, Foz Sul (FS) and Foz Norte (FN) (Fig. 1). Water levels in the flood plain are very close (tens of centimetres) to river water levels and the annual fluctuations in the levels of these lakes are almost synchronous with river fluctuations. Maximum and minimum levels occur during May–June and November–December, respectively. The residence time of water in the várzea is approximately 4–5 months.

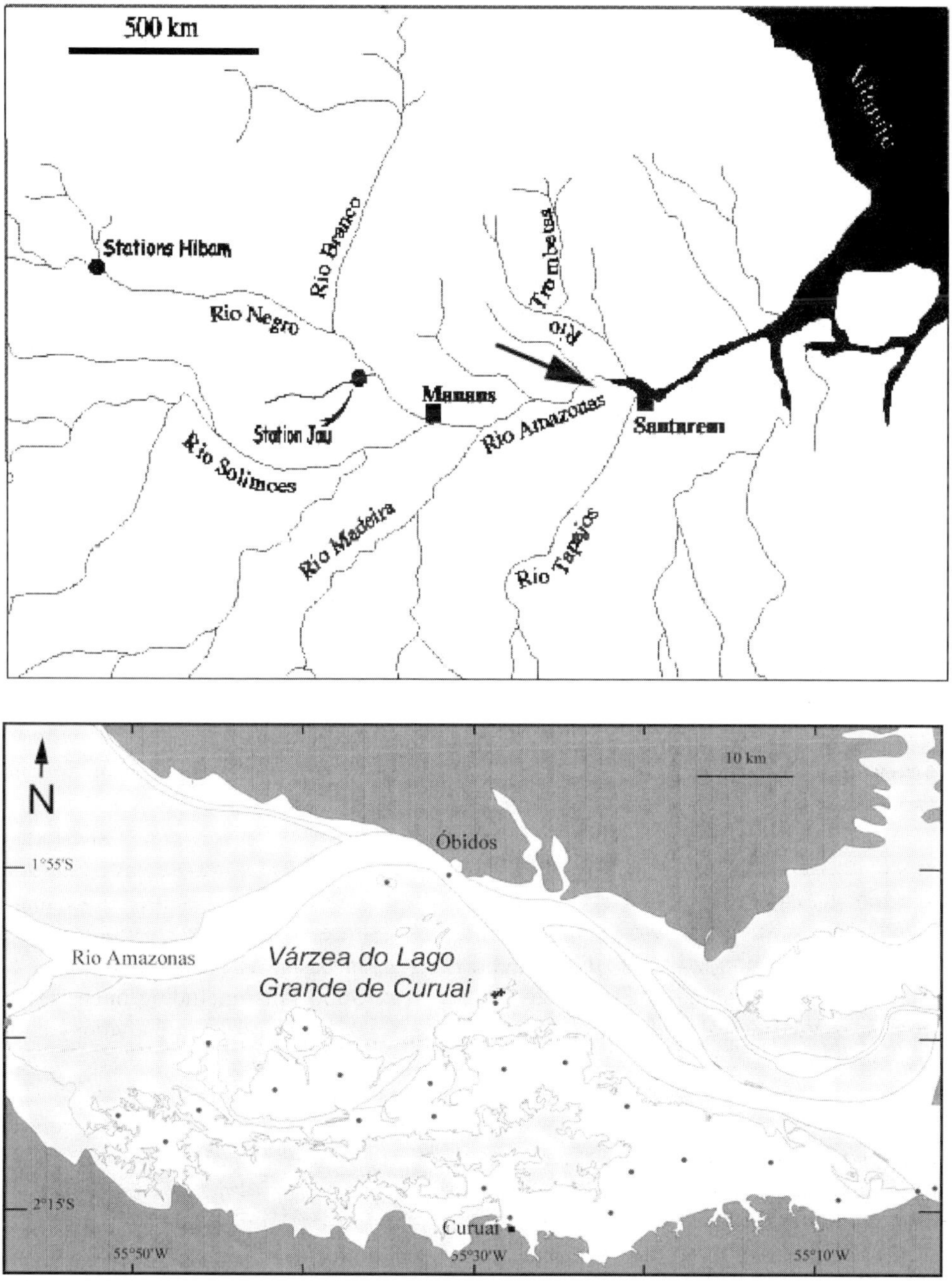

Fig. 1 Location of Várzea Grande de Curuai in the right margin of the Amazon River near Obidos city. Sampling stations are represented with •.

MATERIALS AND METHODS

Water samples were collected along the different lakes in the várzea and the Amazon River (Fig. 1) during seven different cruises designed to sample different water stage periods (Fig. 2). Water discharge was measured at each input and output channel in the várzea with an Acoustic Doppler Current Profiler (RDI Instruments), and at the Obidos station (the last gauging station in the Amazon River). Supporting parameters (pH, conductivity and alkalinity) were measured onboard, using routine conventional techniques. Water samples were analysed onboard, approximately 1 litre of water was filtered (0.7 μm) to separate particulate organic matter (particulate organic carbon (POC) and particulate organic nitrogen (PON)), from dissolved organic matter under reduced pressure, with an all-glass filter holder (Millipore), on 47 mm pre-weighed glass fibre filters (Whatman GF/F), pre-combusted overnight at 450°C. After filtration, filters were dried for 24 h at 50°C. Particulate organic carbon (POC) and particulate organic nitrogen (PON) were analysed by an elemental C/H/N FISONS NA-2000. Dissolved organic carbon (DOC) samples were acidified onboard with ultrapure H_3PO_4 and analysed in the laboratory by a high-temperature catalytic oxidation method (HTCO), using a Shimadzu TOC-5000 instrument. Instantaneous carbon fluxes of each fraction (DOC, POC TOC) were obtained by multiplying the measured water discharges by the corresponding organic carbon concentration at each channel.

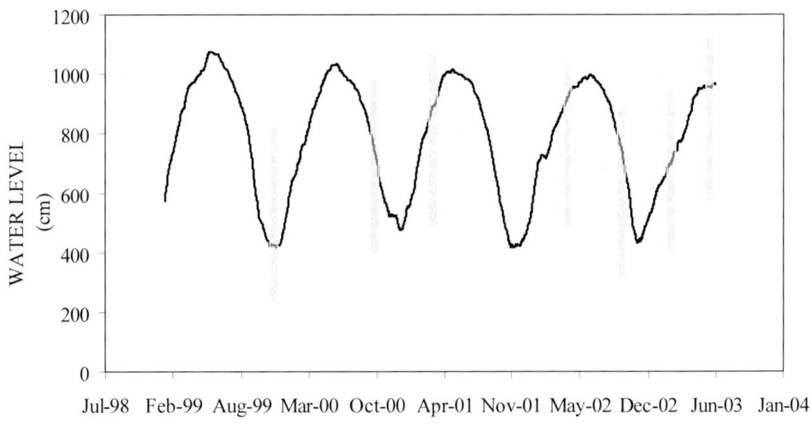

Fig. 2 Time of the different sampling surveys on the hydrograph of the Várzea Grande de Curuai.

RESULTS AND DISCUSSION

Hydrological system

The sample periods were chosen to characterize the different water stages (Fig. 2) in the flood plain system to represent low, rising, falling and high water periods. The cruise undertaken in November 1999 was typical of the low water level (approx. 4 m) in flood plain lakes. The October 2000 and the October 2002 cruises characterize the falling water period (with a water level range between 5.0 and 6.0 m at Curuai station). The April 2001 (approx. 9.25 m at Cu station), March 2002 and March 2003 cruises were characteristics of a rising water period (with a water level range between 7.5 and 9.5 m at Curuai station). The cruise undertaken in June 2003 represents the maximum water level (approx. 10 m) in the flood plain. In exceptional years the water level in the flood plain can reach 11 m. In general, the input of Amazon waters into várzea lakes occurs between January and April and the export of water from várzea waters to the Amazon River takes place during May to December. During rising and highwater levels, the tidal influence over the discharge from the flood plain to the Amazon River cannot be observed. During falling and low water periods it is possible to observe the tidal influence in the flood plain lakes (Kosuth, 2002). The annual water cycle in the várzea is simultaneous with that of the Amazon River.

Total suspended solids and organic matter

Total suspended solids (TSS) and organic matter (dissolved and particulate) showed large seasonal variations. In the várzea, lower concentrations of TSS were found during the rising periods while higher concentrations were found in the Amazon reference site (Obidos station). Conversely, the inverse was recorded during falling water periods. TSS was higher in the várzea lake and lower in the Amazon River. The same behaviour (higher concentrations with falling water stage) was found for organic carbon and nitrogen. Generally, the organic carbon content was inversely correlated with TSS (Fig. 3), mainly due to sediment remobilization. Moreira-Turcq *et al.* (2004) found lower sediment fluxes from water to sediment due to an increase of water velocity, which blocked sedimentation.

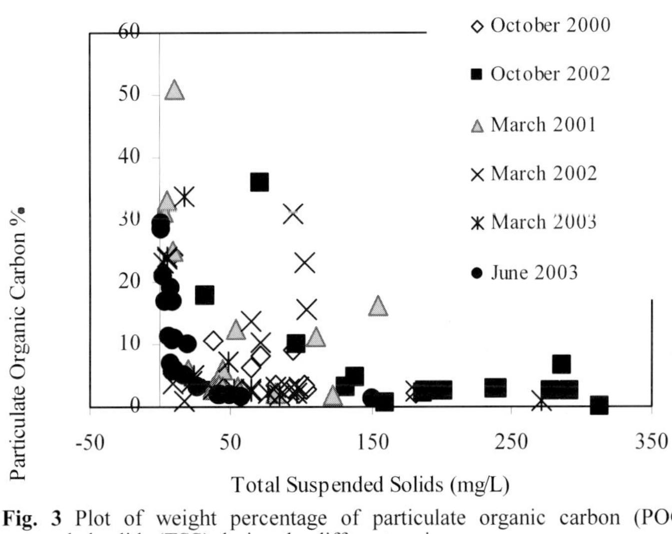

Fig. 3 Plot of weight percentage of particulate organic carbon (POC) *vs* total suspended solids (TSS) during the different cruises.

During the period of study, organic carbon concentrations were very variable and appeared to be very dependent on *in situ* production and water “colour” composition (black and white waters). Traditionally, Amazon tributaries and flood plain lakes are classified according to their colour, which is a function of the relative dominance of dissolved organic matter (blackwater tributaries), or suspended sediment (whitewater tributaries) (Stallard & Edmond, 1983). Organic carbon contents in the Várzea do Lago Grande de Curuai varied from 2 to 30% of the total suspended matter, but may reach 50% during extensive phytoplankton blooms that occur in the white water lakes, mainly during rising periods. This is not described as seasonal behaviour as blooms are observed all the time and seem to be driven by a contribution of carbon for suspended sediments and buried sediments (Moreira-Turcq *et al.*, 2004) that can reach 100 g C m^{-2} $year^{-1}$ in the “Curuai” and are characterized by a terrestrial origin.

In the Amazon River reference (Obidos station), particulate organic carbon concentration is very constant and represents approx. 1–2% of the total suspended matter during the different water seasons.

The C/N ratio (Table 1), based on the atomic ratio, allows the nature (terrestrial or aquatic) of organic matter present in the system to be identified. In flood plain lakes, an interesting feature seems to demonstrate the presence of two types of organic matter during the two different water level stages. During the rising water period the organic matter is characterized by a higher C/N (approx. 9.5) which seems to indicate the organic matter has a terrestrial origin characteristic of the Amazon Basin. Suspended matter present in the Amazon River shows a mean C/N (approx. 9) similar to that recorded for soils (Hedges *et al.*, 1986). The low C/N observed in the flood plain samples can be attributed to a presence of phytoplankton and periphytoplankton blooms (C/N = 6.8 and 7.1) and bacteria activity (C/N = 4.5), respectively. On the other hand, the low C/N (between 4 and 7) found during the falling water period seems to demonstrate an aquatic origin. While bacteria and plankton populations present a very low range of atomic C/N varying between 4–5/1 and 6–7/1, respectively (Amon & Benner, 1996). Dissolved organic carbon concentration has a wide range (5–30 mg l^{-1}) in flood plain lakes, while in the Amazon River its concentration is between 3 and 5 mg l^{-1}. The mean DOC in flood plain lakes is approx. 5.5 mg l^{-1}. The highest concentrations were found during falling and high water levels. A high variability is found between the different lakes and during the different seasons.

Fluxes of organic carbon

Water and matter exchange between flood plain lakes and the Amazon River are highly dynamic. Large quantities of organic carbon are exchanged between the two systems and drive different biogeochemical processes such as adsorption of DOC onto particulate clays. Total Organic Carbon (TOC) in the Amazon River is composed of approx. 70–90% of DOC. In spite of this DOC is generally the main component of Total Organic Carbon (TOC) in flood plain lakes, during low and falling water periods but POC represents an important fraction exported to the river (Table 1). The organic matter which is exported is characterized by a low C/N ratio (6–7) implying a more labile and fresh organic matter than that present in the river. Carbon exchange (import

Table 1 Water level, net balance, discharge, dissolved organic carbon (DOC) flux, particulate organic carbon (POC) flux, total organic carbon (TOC) flux, DOC/TOC ratio and C/N ratio during annuals cycle in the flood plain "Várzea do lago grande de Curuai".

Cruise	Water level	Balance	Discharge ($m^3 s^{-1}$)	DOC flux (kg C s^{-1})	POC flux (kg C s^{-1})	TOC flux (kg C s^{-1})	DOC/TOC (%)	C/N ratio
November 1999	low	export	81	0.397	1.282	1.679	24	7
October 2000	falling	export	950	8.452	3.91	12.36	68	5
March 2001	rising	import	847	9.78	1.81	11.59	84	10
March 2002	rising	import	200–284	1.21	0.43	1.64	74	9
October 2002	falling	export	755	3.424	4.327	7.751	44	7
March 2003	rising	import	300	1.08	0.693	1.77	61	10
June 2003	high	export	1062	5.69	0.741	6.42	89	8

and/or export) reaches 11–12 kg C s^{-1}. Export of organic carbon to the river occurs during falling water periods and varied between 1.68 and 12.36 kg C s^{-1}. During rising periods the river organic carbon flux to lakes is also great (1.64–11.59 kg C s^{-1}), with the dissolved organic carbon being the main component.

CONCLUSIONS

Exchanges between flood plain lakes and the Amazon River are always present. Flood plain lakes receive organic matter from the river for six months and also supply the Amazon River with organic matter in the remaining months. The import and export of organic carbon between the Amazon and the flood plain seem to be in the same order, and dissolved organic carbon is the main fraction. The more important feature is that the nature of the organic matter imported and that exported seems to differ. Organic matter that is imported is more refractory while organic matter that is exported seems to be more labile, and consequently more reactive.

Acknowledgements This research was supported by the joint CNPq (Conselho Nacional de Desenvolvimento e Pesquisa Tecnologica)–IRD (Institut de Recherche pour le Developpement) project HIBAm (Hydrology and Geochemistry of Amazon Basin). We would like to thank the Agencia Nacional das Aguas (ANA) and Companhia de Pesquisa dos Recursos Minerais (CPRM) technical groups for their help during the boat cruises.

REFERENCES

Amon, R. M. W. & Benner, R. (1996) Bacterial utilization of different size classes of dissolved organic matter. *Limnol. Oceanogr.* **41**(1), 41–51.

Bartlett, K. B. & Hariss, R. C. (1993) Review and assessment of methane emissions from wetlands. *Chemosphere* **26**(1–4), 261–320.

Dunne, T., Mertes, L. A., Meade, R. H., Richey, J. E. & Forsberg, B. R. (1998) Exchanges of sediment between the flood plain and channel of the Amazon River in Brazil. *GSA Bull.* **110**(4), 450–467.

Ertel, J. R., Hedges, J. I., Devol, A. H., Richey, J. E. & Ribeiro, M. N. G. (1986) Dissolved humic substances of the Amazon River system. *Limnol. Oceanogr.* **31**(4), 739–754.

Filizola, N. (1999) O fluxo de sedimentos em suspensão nos rios da Bacia Amazônica brasileira. Publ. ANEEL (Agencia Nacional de Energia Eletrica), Brasilia, Brazil.

Hedges, J. I., Clark, W. A. Quay, P. D., Richey, J. E., Devol, A. & Santos, U. (1986) Composition and fluxes of particulate organic material in the Amazon River. *Limnol. Oceanogr.* **31**(4), 717–738.

Irion, G., Keim, G., Muller J., Nunes de Mello, J. & Junk, W. J. (1999) The Late Quaternary river and lake development in Central Amazonia. In: *Hydrological and Geochemical Processes in Large Scale River Basins* (Int. Symp., Manaus, Brazil, 15–19 November).

Junk, W. J. (1997) *The Central Amazon Flood Plain: Ecology of a Pulsing System.* Ecological Studies 126. Springer Verlag Berlin, Germany.

Kosuth, P. (2002) A case study of flood plains dynamics: dynamics, water balance, sediment balance of the Várzea of Lago Grande de Curuai. In: *Second LBA Conference* (Manaus, Brazil, 7–10 July).

Melack, J. M. & Fisher, T. R. (1990) Comparative limnology of tropical flood plain lakes with an emphasis on the central Amazon. *Acta Limnol. Brasil.* **3**, 1–48.

Melack, J. M. & Forsberg, B. R. (2001) Biogeochemistry of Amazon Flood plain Lakes and Associated Wetlands. In: *The Biogeochemistry of the Amazon Basin* (ed. by M. E. McClain, R. L. Victoria, & J. E. Richey) 235–274. Oxford Univ. Press, New York, USA.

Moreira-Turcq, P. F., Seyler, P., Guyot, J. L. & Etcheber, H. (2003a) Exportation of organic carbon from the Amazon River and its main tributaries. *Hydrol. Processes* **17**, 1329–1344.

Moreira-Turcq, P. F., Seyler, P., Guyot, J. L. & Etcheber, H. (2003b) Characteristics of organic matter in the mixing zone of the Rio Negro and Rio Solimões of the Amazon River. *Hydrol. Processes* **17**, 1393–1404.

Moreira-Turcq, P., Jouanneau, J. M., Turcq, B., Seyler, P., Weber, O. & Guyot, J. L. (2004) Carbon sedimentation at Lago Grande de Curuai, a flood plain lake in the low Amazon region: insights into sedimentation rates. *Palaeogeogr. Palaeoclim. Palaeoecol.* **214**, 27–40.

Quay, P. D., Wilbur, D. O., Richey, J. E., Hedges, J. I., Devol, A. H. & Victoria, R. (1992) Carbon cycling in the Amazon River: implications from the ^{13}C compositions of particles and solutes. *Limnol. Oceanogr.* **37**(4), 857–871.

Richey, J. R. (1982) The Amazon River system: a biogeochemical model. *Mitt. Geol. Paläont. Inst. Univ. Hamburg* **52**, 365–378.

Richey, J. R., Hedges, J. I., Devol, A. H., Quay, P. D., Victoria, R., Martinelli, L. A. & Forsberg, B. R. (1990) Biogeochemistry of carbon in the Amazon River. *Limnol. Oceanogr.* **35**, 352–371.

Richey, J. E., Melack, J. M., Aufdenkampe, A. K., Ballester, V. M. & Hess, L. L. (2002) Outgassing from Amazonian rivers and wetlands as a large tropical source of atmospheric CO_2. *Nature* **416**, 617–620.

Stallard, R. F. & Edmond, J. M. (1987) Geoquemistry of the Amazon: 3 Weathering chemistry and limits to dissolved inputs. *J. Geophys. Res.* **92**, 8293–8302.

2 Nutrients

Nitrogen cycling on five headwater forested catchments in Mid-Appalachians of Pennsylvania

DAVID R. DEWALLE, ANTHONY R. BUDA, JENNIFER A. EISMEIER, WILLIAM E. SHARPE, BRYAN R. SWISTOCK, PATRICIA L. CRAIG & MICHAEL A. O'DRISCOLL

School of Forest Resources and Penn State Institutes of the Environment, 107 Land and Water Research Bldg., Pennsylvania State University, University Park, Pennsylvania 16802, USA
drdewalle@psu.edu

Abstract Nitrogen cycling has been studied since 1988 on five headwater forested catchments in the Mid-Appalachian region of northeast United States to determine impacts of atmospheric deposition. Nitrogen dissolved in streams was dominated by NO_3-N but dissolved organic nitrogen was a significant component of stream export. Watershed input–output budgets showed nitrogen retention varied from 63–96% of estimated atmospheric deposition inputs. Retention and losses of nitrogen occurred primarily in the uplands on Baldwin Creek basin, with lesser losses occurring in riparian lowland regions around seeps and the main stream corridor. With the exception of one basin that experienced forest decline and salvage logging, no significant trends in stream NO_3-N concentrations have been detected over the past 15 years due to the Clean Air Act.

Key words assimilation; denitrification; dissolved organic nitrogen; dry deposition; hyporheic zone; nitrogen budgets; salvage logging; seeps; wet deposition

INTRODUCTION

Headwater forested catchments in the Appalachian Plateaus Province of Pennsylvania play an important role in society by providing domestic and industrial water supplies, protection for native biodiversity, aquatic habitat, and recreational opportunities. These forested catchments and similar basins throughout the northeast United States are threatened by relatively high levels of atmospheric deposition, which have resulted in changes in water and soil chemistry. The Clean Air Act of 1990 (CAA) and subsequent amendments were enacted to reduce levels of sulphur and nitrogen emissions to the atmosphere and prevent watershed acidification. A monitoring network of streams and lakes was established in 1990 (US EPA, 2003), including the five Pennsylvania watersheds reported on in this paper, to determine the effect of the CAA on stream water quality. Although initial emphasis of CAA was placed upon reductions in sulphur deposition, current and future emphasis is being placed upon reductions in nitrogen deposition.

This paper reports on our current understanding of nitrogen cycling on five forested watersheds on the Appalachian Plateaus Province in Pennsylvania, with an emphasis on the role of atmospheric deposition and riparian zone processes. Objectives of the research were to determine the trends in stream chemistry due to the CAA, to determine the overall input–output nitrogen budget of the five catchments as affected by local conditions, to determine the role of uplands *vs* riparian zones in controlling the export of nitrogen, and to determine the significance of dissolved organic nitrogen (DON) as a component of stream nitrogen export.

STUDY AREAS

The locations of the five watersheds, Baldwin Creek and Linn Run in southwest Pennsylvania and Benner, Stone and Roberts Runs in northcentral Pennsylvania, are shown in Fig. 1. All basins are second order with areas of about 1100 ha, except for Baldwin Creek, which has an area of about 535 ha. Rocks on these unglaciated basins are largely sedimentary sandstones, shales, limestones and conglomerates. Residual soils are stony, sandy loams. Average basin slopes are about 7–8%, except on the smaller Baldwin Creek basin that has an average slope of 15%. Watersheds are entirely forested with mixed deciduous forest and, except for Baldwin Creek, have not experienced significant recent cutting. Salvage logging was conducted on about 60% of Baldwin Creek basin in 1990–1992 due to forest decline and nitrogen levels in the stream have since returned to approximate pre-cut levels. The Baldwin Creek watershed has suffered forest decline at higher elevations for approximately 20 years presumably as a consequence of acid deposition. Dead and dying northern red oak (*Quercus rubra*) were harvested from 340 ha of the 535 ha watershed in the period 1990–1992. Harvests were both by selection cut (114 ha) and clear-cut (226 ha). Woody regeneration (<0.03 stems m^{-2}) was very poor throughout the watershed, and the remaining 40% of overstory trees were either dead or had >50% crown loss in June 2000. Precipitation on these basins averages about 100–120 cm annually and mean annual air temperatures average about 7–8°C. Levels of acidic atmospheric deposition within the region are high and wet and dry deposition data from monitoring stations that are part of the US National Atmospheric Deposition Program (NADP 2004), PA Department of Environmental Protection Acid Rain network (PA DEP 2004), and Clean Air Status and Trends Network (CASTNET 2004) are available (see Fig. 1).

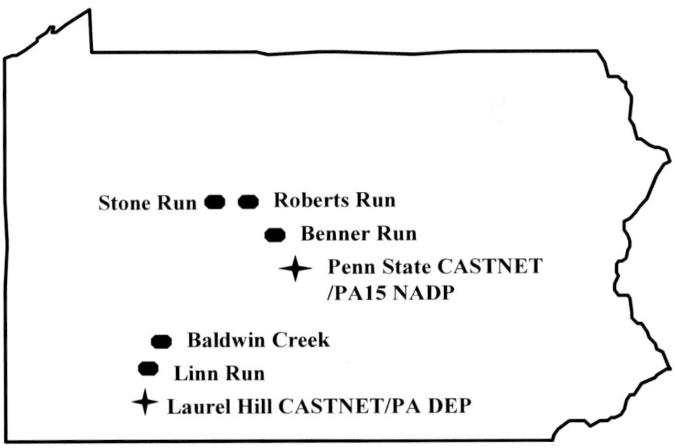

Fig. 1 Map of Pennsylvania, USA, showing locations of five forested watersheds and reference wet and dry atmospheric deposition stations.

METHODS

On each watershed streamflow has been monitored using continuous stage recorders and natural channels for control since 1988. Watershed runoff averages about 40–60 cm annually. Initially stream chemistry was intensively studied during acidic runoff episodes (DeWalle & Swistock, 1994; Wigington *et al.*, 1996), but since about 1990 streams have been sampled only monthly, with a few interruptions, mostly at baseflow. Water samples from streams were analysed for NO_3-N concentrations by ion chromatography, NH_4-N by the phenate method, and total dissolved nitrogen (TDN) and total nitrogen (TN) by the persulphate oxidation method (APHA, 1998). Dissolved organic nitrogen (DON) concentrations were determined as DON = TDN – (NO_3-N + NH_4-N). In a limited set of baseflow samples particulate nitrogen concentrations were also evaluated as the difference between TN and TDN. Export of nitrogen in streamflow was computed as the period-weighted product of flow rates and concentrations for comparison with computed inputs in atmospheric deposition.

RESULTS

Nitrogen in streamflow

The importance of dissolved organic nitrogen and other ions in streamflow was assessed for the 1997–2003 study period. Median TDN and NO_3-N concentrations varied considerably among the five study streams, but NH_4-N concentrations were near or below the detection limit (0.013 mg l^{-1}) for all of the streams. Median DON concentrations for the 7-year period were near 0.04 mg l^{-1} on four streams and 0.02 mg l^{-1} on Stone Run (Fig. 2). DON comprised 10–30% of TDN on these five streams (Fig. 2). Particulate nitrogen concentrations were very low (median 0 to 0.01 mg l^{-1}) in all five streams based upon 20 mostly low flow samples.

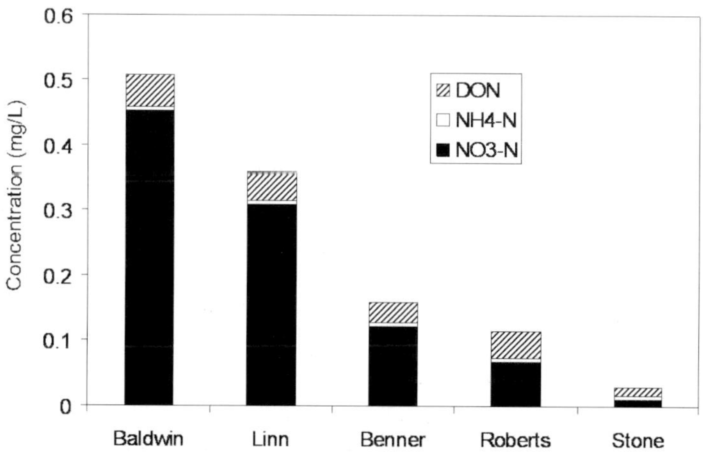

Fig. 2 Median concentrations of NO_3-N, NH_4-N and DON in each of the five streams in Pennsylvania from 1997–2003.

DON concentrations for baseflow samples were generally stable on these streams with surprisingly little variability or correlation with flow or other nitrogen species. In fact, DON was not significantly correlated with NO_3-N on any of the study streams (P = 0.33 to 0.78). Separating the data into seasons did not improve the relationship. DON was not significantly correlated with stream flow, although the data set is limited to fixed interval, monthly stream samples representing low to moderate flow conditions.

A separate study examining nitrogen export that included storm flow sampling was conducted on Benner Run from October 1999 to October 2000 (Craig, 2001). The analyses included bimonthly base flow samples and samples from six storms: three autumn storms and three spring storms. For the 1999–2000 study, the median concentration of NO_3-N was 0.099 mg l^{-1}, which was similar to the 7-year median NO_3-N concentration of 0.121 mg l^{-1}. However, the median DON concentration of 0.096 mg l^{-1} was significantly higher than the 7-year median DON concentration of 0.032 mg l^{-1}. As in the 7-year data, no concentration–discharge relationship was evident for DON. However, when data were grouped by flow type (rising, peak and falling storm flows), differences in median concentrations among flow types were significant except for NH_4-N. NO_3-N concentrations were highest during base flow and lowest during the falling-limb flow, whereas DON concentrations were highest during falling-limb flow and lowest during base flow conditions. DOC concentrations were highest during peak flow and lowest during base flow. The higher DON concentrations measured during elevated flow may explain the higher median DON concentration observed during this study as compared to the 7-year data.

Nitrogen input–output budgets

Nitrogen input–output budgets for the five watersheds are presented in Fig. 3 for the period October 1997–September 2001, extending previous work conducted by Sweeney (1998) and Dow & DeWalle (1997). The mean annual total nitrogen deposition for Laurel Hill was about 9.7 kg ha^{-1}. At the Penn State site in northcentral Pennsylvania, mean annual total nitrogen deposition was only 7.9 kg ha^{-1}. Dry nitrogen deposition contributed about 25% of the total observed nitrogen deposition at Laurel Hill and about 47% of the total observed nitrogen deposition at Penn State.

Average stream export of N was highest for Baldwin Creek (3.6 kg ha^{-1}) followed by Linn Run (2.1 kg ha^{-1}), Benner Run (1.7 kg ha^{-1}), Roberts Run (1.0 kg ha^{-1}), and Stone Run (0.3 kg ha^{-1}). Between 1990 and 1997 stream NO_3-N levels on Baldwin Creek basin were elevated by forest decline and salvage logging which influenced the export average for this basin. Differences between export and atmospheric inputs (Fig. 3) suggest that total N retention by these forest ecosystems ranges from 6.1 to 7.6 kg ha^{-1} each year. Basin N retention averaged about 63% (Baldwin) and 78% (Linn) of N inputs in southwestern Pennsylvania and about 78–96% of N inputs in northcentral Pennsylvania. Most of the variation among basins was due to variations in dissolved NO_3-N export. These results illustrate the relatively high retention of atmospheric deposition N inputs typical of forested basins in the region (DeWalle & Pionke, 1996), but also relatively high variation in absolute levels of stream N export.

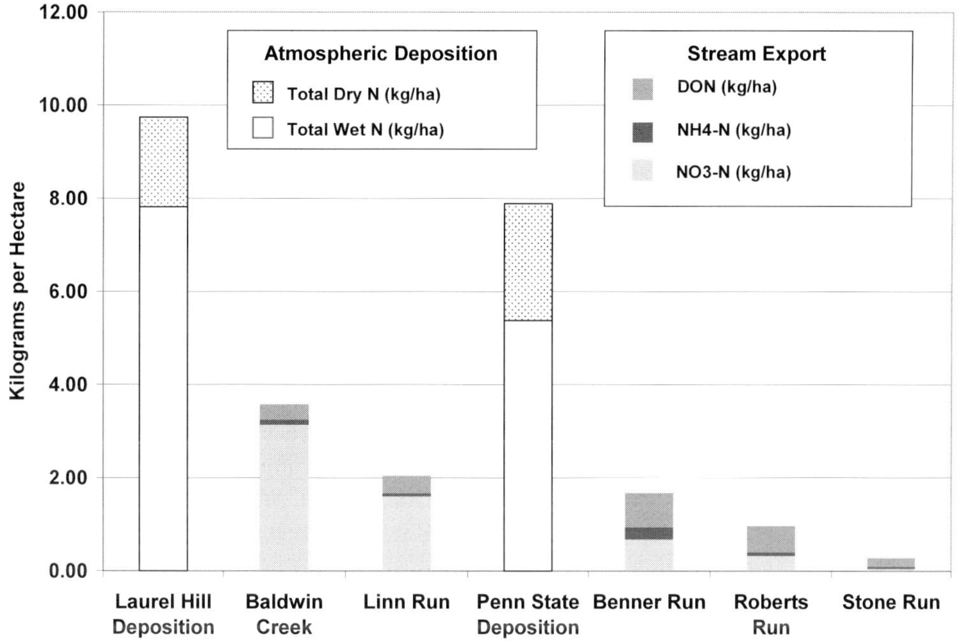

Fig. 3 Average annual nitrogen input–output budgets for five forested Appalachian watersheds during water years 1997–2001.

Riparian zone nitrogen cycling

To better understand the relative significance of upland versus riparian zone N retention, a N mass balance study was conducted on Baldwin Creek from May 2002 to April 2003. Measurements of stream flow rates and sampling for TDN in the stream and at the head and base of two major seep or spring runs within each of seven longitudinal stream sections on the basin were made on one random day per month. Measurements were used to compute the TDN biogeochemical loss or retention from the riparian zone (*TDN Loss*) in kg day^{-1} for each stream section as:

$$TDN\ Loss = [(Q_{dn} \times C_{dn}) - (Q_{up} \times C_{up})] - [(Q_{dn} - Q_{up}) \times C_{gw}] - (Q_{trib} \times C_{trib})$$

where Q is discharge in m^3 day^{-1}, C is TDN concentration kg m^{-3}, *up* denotes upstream station, *dn* denotes downstream station, *gw* denotes groundwater, *trib* denotes any tributary stream. Only one intermittent tributary occurred on the basin. The change in flow in each stream section ($Q_{dn} - Q_{dup}$) was assumed to be due to groundwater inputs or losses plus tributary inputs. Sampled seeps and springs which appeared as discrete surface flows to the main channel were assumed to represent groundwater TDN, since piezometer head data show the main channel was losing water. *TDN Loss* was summed for all seven stream sections to obtain a total riparian zone loss on each day. Use of TDN concentrations from the base of seeps and springs for C_{gw}, rather than the TDN at the head of seeps or springs, allowed separate calculation of losses from main stream segments only. Daily *TDN Loss* data from each month varied with flow rate and these data were scaled to obtain an annual loss estimate using the frequency distribution of daily flow rates in that year. N inputs due to fixation of atmospheric N and N losses due to stream export of particulate N were assumed negligible.

The annual average atmospheric inputs of N for this catchment were estimated to be 9.7 kg ha^{-1} at the Laurel Hill site. TDN export in Baldwin Creek streamwater during the one year mass balance study was 3.1 kg ha^{-1}. Total loss or retention of TDN on the watershed for the year was computed as the inputs minus export or 9.7 – 3.1 = 6.6 kg ha^{-1} (Fig. 4). Total loss or retention from the combined seepage zones and main stream corridor for the year was estimated to be 1.6 kg ha^{-1}, which represented 24% of total watershed TDN loss or retention. Thus, 76% of total watershed TDN loss could be attributed to retention or loss from the uplands portion of the catchment. The total 24% riparian loss could be further partitioned between losses in seepage zones (14%, 0.92 kg ha^{-1}) and losses in the main stream corridor (10%, 0.67 kg ha^{-1}). Seepage and spring run zones had greater N retention or losses than the main channel region. The processes controlling upland N losses on Baldwin Creek are attributed to assimilation by upland vegetation and soil organisms, as evidenced by the substantial increase in NO_3-N following salvage logging on Baldwin Creek basin. Within the riparian zone, losses are attributed to assimilation by the aquatic ecosystem and denitrification.

Forest decline and salvage logging

Both NO_3-N concentrations and export increased sharply following salvage harvest and remained elevated throughout 1998 (Herrmann *et al.*, 2001). Nitrate-N export

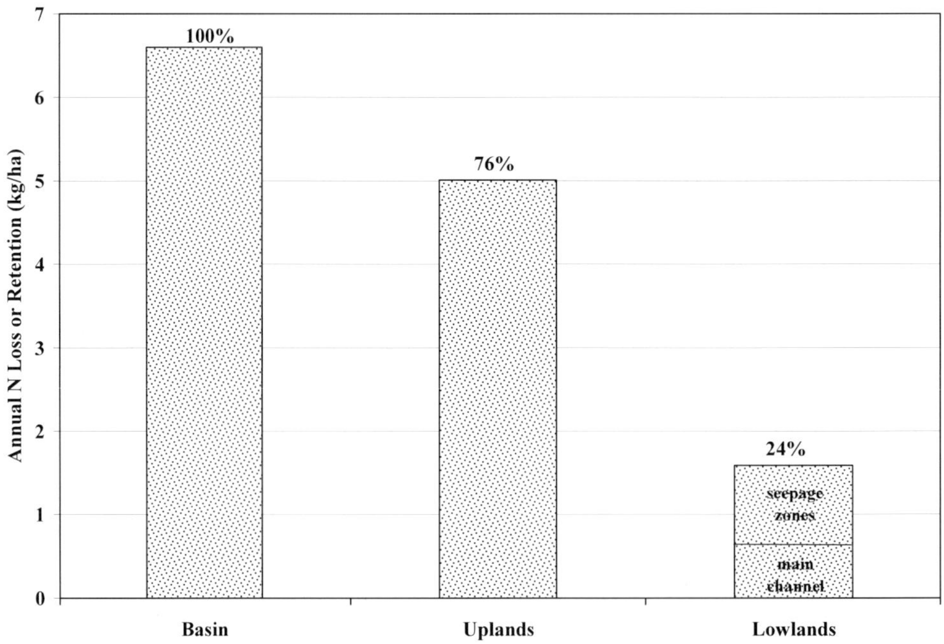

Fig. 4 Estimated Annual Total, Uplands and Lowlands Losses or Retention of N on Baldwin Creek Basin, May 2002–April 2003.

was highest in 1993 and 1994 and averaged 15.5 kg ha^{-1} of harvested forest; a yield that far exceeded all other regional estimates of nitrate loss in the first two years post-cutting (0.1 to 4.5 kg ha^{-1}), except that observed at Hubbard Brook in New Hampshire (248 kg ha^{-1}). These data suggest that logging in areas with forest decline may cause levels of NO_3-N in streams greater than expected from logging healthy forests.

Long-term nitrogen trends

Since the passage of the 1990 Clean Air Act Amendments, SO_4 and NO_3-N concentrations and deposition (kg ha^{-1}) in precipitation have significantly declined in Pennsylvania. (Lynch *et al.*, 2003). Although SO_4 concentrations have significantly declined in most streams, concentrations of NO_3-N in streams have shown non-significant declines in Stone, Roberts and Benner Runs. Figure 5 shows trends in stream SO_4 and NO_3-N concentrations (mg l^{-1}) for Roberts Run watershed from 1988 to 2002.

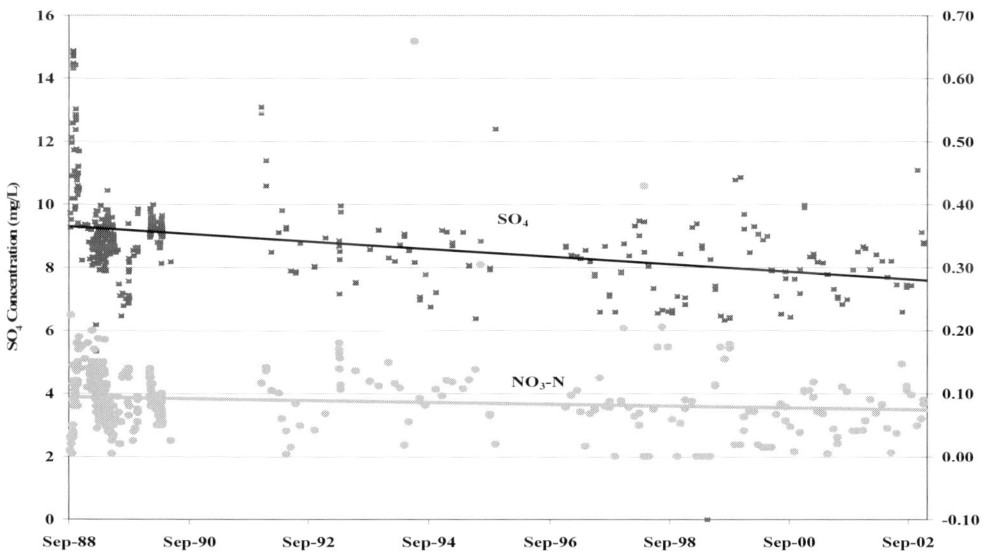

Fig. 5 Trends in stream water SO_4 and NO_3-N concentrations (mg l^{-1}) for Roberts Run watershed (1988–2002).

CONCLUSIONS

The five Appalachian forested watersheds retain the majority of nitrogen from atmospheric wet and dry deposition, but wide variations exist (63–96% of deposition) in retention efficiency. Upland forest is primarily responsible for N losses, but lowland riparian zones also play a significant role. Off-channel seepage and spring zones appeared more important than the main channel region in controlling N losses and retention on Baldwin Creek. Nitrate levels in streams have not shown a significant decline due to the Clean Air Act, although slight negative trends are evident. Dissolved organic nitrogen should be included as a component of stream export especially where stream nitrate levels are low.

Acknowledgements The support of this research by Penn State University, Institutes of the Environment and by the US Environmental Protection Agency is gratefully acknowledged. Cooperation of the Pennsylvania Game Commission and PA Dept. Conservation and Natural Resources in providing access to watersheds is appreciated.

REFERENCES

APHA (1998) *Standard Methods for the Examination of Water and Wastewater* (20th edn). Amer. Public Health Assoc., Washington, DC, USA.

CASTNET (2004) *Clean Air Trends and Status Network*. United States Environmental Protection Agency, Washington, DC, USA. http://www.epa.gov/castnet/sites.html

Craig, P. L. (2001) Significance of organic and particulate nitrogen in an Appalachian forest stream. MS Thesis, Environ. Poll. Control, Pennsylvania State University, Pennsylvania, USA.

DeWalle, D. R. & Swistock, B. R. (1994) Causes of episodic acidification in five Pennsylvania streams on the northern Appalachian Plateau. *Water Resour. Res.* **30**(7), 1955–1963.

DeWalle, D. R. & Pionke, H. B. (1996) Nitrogen export from forest land in the Chesapeake Bay region. In: *Proc. 1994 Chesapeake Bay Res. Conf.* (Edgewater, Maryland, USA), 649–655. Chesapeake Bay Res. Corp.

Dow, C. L. & DeWalle, D. R. (1997) Sulfur and nitrogen budgets for five forested Appalachian Plateau Basins. *Hydrol. Processes* **11**, 801–816.

Herrmann, M., Sharpe, W. E., DeWalle, D. R. & Swistock, B. R. (2001) Nitrogen export from a watershed subjected to partial salvage logging. *Scientific World J.* **1**(11–S2), 440–448.

Lynch, J. A., Horner, K. S. & Grimm, J. W. (2003) Atmospheric deposition: spatial and temporal variations in Pennsylvania-2002. Penn State Univ., Institutes of the Environment, University Park, PA. Dec. 2003. also available at http://www.dep.state.us/dep/deputate/airwaste/aq/acidrain/2002rep/02_report.htm

NADP (2004) National Atmospheric Deposition Program Office, Illinois State Water Survey. Champaign, Illinois, USA. http://nadp.sws.uiuc.edu/

PA DEP (2004) Acid Rain and Mercury Monitoring Sites. Pennsylvania Dept. Environ. Protection, Pennsylvania, USA. http://www.dep.state.us/dep/deputate/airwaste/aq/acidrain/sitemap/htm

Sweeney, J. S. (1998) Chemical mass balances for five Appalachian Plateau watersheds in Pennsylvania (1988–1995). Master of Engng in Environ. Poll. Control Paper, Pennsylvania State University, Pennsylvania, USA.

US Environmental Protection Agency (2003) Response of surface water chemistry to the clean air act amendments of 1990. EPA 620/R-03/001.

Wigington, P. J., Baker, J. P, DeWalle, D. R., Kretser, W. A., Murdoch, P. S., Simonin, H. A., Van Sickle, J., McDowell, M. K., Peck, D. V. & Barchet W. R. (1996) Episodic acidification of small streams in the Northeastern United States: Episoidic Response Project. *Ecol. Applic.* **6**(2), 374–388.

The exchange of phosphorus between riparian wetland sediments, pore water and surface water

BEN SURRIDGE[1], LOUISE HEATHWAITE[2] & ANDREW BAIRD[3]

1 *Catchment Science Centre, Groundwater Protection and Restoration Group, Department of Civil and Structural Engineering, University of Sheffield, Sir Frederick Mappin Building, Mappin Street, Sheffield S1 3JD, UK*
b.surridge@sheffield.ac.uk

2 *The Lancaster Environment Centre, Lancaster University, Bailrigg, Lancaster LA1 4YQ, UK*

3 *Department of Geography, Queen Mary, University of London, Mile End, London E1 4NS, UK*

Abstract Phosphorus (P) is generally viewed as the main element driving eutrophication risk in freshwaters. In this paper we examine the release of P from riparian wetland sediments to pore water and surface water. The research was conducted at Strumpshaw Fen in the Norfolk Broads region of the UK. We show that substantial quantities of bioavailable P may be released into solution from the sediments at Strumpshaw Fen. This release is physico-chemically controlled, and involves the reductive-dissolution of P complexed at the surface of ferric hydroxide. Maximum concentrations of molybdate-reactive P (MRP) in 0.45 µm filtered solutions can reach up to 1 mg l^{-1} P in surface water, and up to 4.5 mg l^{-1} P in pore water. Such high concentrations of MRP indicate that riparian wetlands may act as sources of P with the potential to threaten the chemical and ecological quality of receiving water bodies.

Key words eutrophication; iron; Norfolk Broads, UK; phosphorus; riparian wetland

INTRODUCTION

Within catchments, wetlands are unique ecosystems in the continuum of terrestrial to aquatic environments (Aerts *et al.*, 1999), often at the interface between upland areas and receiving water bodies (Reddy *et al.*, 1999; Prior & Johnes, 2002). This interface has become a focus for academic and applied interest, due to the potential for wetlands to mediate anthropic impacts from upland regions before they reach economically and environmentally valuable receiving waters (Burt *et al.*, 2002). Wetlands may regulate the dissolved pollutant (including nutrient) and suspended sediment loads delivered to adjacent rivers, lakes and coastal waters. Therefore, wetlands could represent an important catchment element contributing to the improvements in the chemical and ecological quality of freshwaters required by legislative drivers such as the European Union Water Framework Directive (European Parliament, 2000).

One of the most important criteria of the quality of freshwaters is nutrient status, owing to the consequences of excess nutrients in terms of eutrophication. Phosphorus (P) is generally viewed as the main element driving eutrophication risk. Wetlands may represent net sources or sinks of P, depending on the quantity and form of P already present in the system, and the input to the system via inflowing waters (Pant *et al.*, 2002; Pant & Reddy, 2003; Soto-Jimenez *et al.*, 2003; Novak *et al.*, 2004). Many flood plains, including those containing riparian wetlands, may have accumulated large stores of sediment-P historically due to their connection with adjacent channel systems carrying significant nutrient loads in dissolved and particulate forms (e.g. Venterink *et al*., 2003; Owens & Walling, 2002). The release of P from riparian wetland sediments may result in the establishment and maintenance of a nutrient-enriched status within the wetland itself. Indeed, in some cases the internal supply of nutrients may even exceed current external anthropic loadings (Fisher & Reddy, 2001), particularly as environmental management often focuses on the reduction of external nutrient loads. Further, under conditions of increased internal nutrient availability, riparian wetlands may become potential source areas for the delivery of P to external receiving water bodies, particularly in systems where hydrological evidence indicates a strong potential for water flow through the wetland sediments.

Despite the potential importance of nutrient release from wetland deposits, little research has focused on the impact of wetland sediments on the chemical quality of the overlying surface water (Fisher & Reddy, 2001). More specifically, there is only very sparse information on the impact of the flooding of riparian soils on the various sediment-P fractions, and consequently on the retention/release of phosphorus (Wright *et al*., 2001; Novak *et al*., 2004). Further, the few studies that have been conducted in wetland systems have dealt almost exclusively with P dynamics within the overlying surface water (e.g. Newman & Pietro, 2001; Pant & Reddy, 2003; White *et al*., 2004). The concentration solely on surface water neglects the potential importance of pore-water P dynamics. The release of P into overlying surface waters will be partly dependent on the phosphorus dynamics within the pore water. Further, given sub-surface hydrological connection, it is possible that P could be delivered to receiving water bodies by direct exchanges with pore water. Such exchanges may alter the nutrient status of a receiving water body without the prior need for phosphorus release to overlying surface water.

The aim of the work reported in this paper was to examine the exchange of phosphorus between a typical riparian wetland sediment and the overlying water column. Exchange between the sediment and pore water was also investigated, in terms of both the impact on P release to the overlying water column, and the concentration of P in the pore water itself. The research was designed to examine the processes that control these exchanges, the rates at which they operate, and whether riparian wetland sediments may act as potential sources for the delivery of phosphorus to receiving water bodies.

SITE DESCRIPTION AND METHODOLOGY

The research reported in this paper was conducted at Strumpshaw Fen, part of the Yare Broads and Marshes Site of Special Scientific Interest in the Norfolk Broads region of the UK. The Broads represent an internationally recognized wetland complex, with some 5500 hectares of Broadland protected under Ramsar designation. The site is located at 52°37′N, 1°26′E in the northern flood plain of the River Yare, approximately 10 km to the east of Norwich. The wetland complex at Strumpshaw comprises 197 hectares of open water, reedbed, carr woodland and grazing marsh within a flood plain that is generally between 500 and 1000 m wide on either side of the river (Fig. 1).

Fig. 1 The wetland complex at Strumpshaw Fen, looking southeast. Photograph courtesy of Mike Page, RSPB (Royal Society for the Protection of Birds) ©RSPB

The fen vegetation is dominated by a *Phragmites australis* Cav. (Streud.). – *Peucedanum palustre* (L.) Moench. mixed fen community. Peat deposits reach depths of 5–6 m in the vicinity of the field site and consist of layers of mixed fen and woody peats, with the woody peats dominating in the lower part of the profile. The wetland complex has been exposed to substantial nutrient loading in the past as a result of over-bank and sub-surface inputs of water from the nutrient-enriched River Yare and a tributary stream the Lackford Run (Environment Agency, unpublished data). An important conduit for the two-way exchange of water between the wetland and the adjacent channel systems is the ditch network that runs throughout Strumpshaw Fen, shown in Fig. 1.

Laboratory core incubations examining the release of P to surface water and pore water following flooding were conducted using intact cores of the surface 50 cm of peat collected from Strumpshaw Fen. This zone of peat encompasses the seasonal water table range (maximum summer drawdown to approximately 40 cm below the peat surface), and is the focus for exchanges of water between the peat and the adjacent ditch network. Therefore the availability of P in this zone was of particular interest. Cores were collected from the field during the summer drawdown period when water tables were approximately 40 cm below the ground surface. Two separate experiments were conducted using two different solutions to re-flood the intact cores, and in each experiment duplicate cores were used. In Experiment A the cores were flooded with de-ionized water, to act as a "control" and to examine the release of P in the absence of external loads of phosphorus or electron acceptors. In Experiment B, simulated River Yare water was used as the flooding solution, made using standard salts dissolved in de-ionized water, with the chemical composition of the solution based on the analysis of samples collected from the River Yare. Following re-flooding of the cores, samples of surface water and pore water were collected, filtered through 0.45 μm membrane filter papers, and analysed for molybdate-reactive phosphorus (MRP) alongside other major ions. Separate intact peat cores were flooded with the solutions used in Experiments A and B described above, and subsequently analysed for sediment-P fractions using the fractionation scheme developed by Reddy *et al.* (1998). The results were compared with those from cores that were not re-flooded, and allowed a direct assessment to be made of the impact of flooding on sediment-bound phosphorus fractions.

RESULTS AND DISCUSSION

The phosphorus content of the sediments at Strumpshaw Fen

Figure 2 shows the distribution of total phosphorus (TP), total organic phosphorus (TP_o) and total inorganic phosphorus (TP_i) in the surface 50 cm of peat at Strumpshaw Fen under drained conditions. The surface 15 cm of peat contained large quantities of TP, with mean values for the three cores exceeding 1200 mg kg^{-1} P. Below 15 cm depth, total phosphorus content declined substantially, until more constant levels between 550 and 650 mg kg^{-1} P were reached and maintained at between 30 and 50 cm depth. Total phosphorus using the phosphorus fractionation scheme was strongly correlated with total phosphorus determined by digestion of fresh soil samples (r = 0.98, $P < 0.01$). Throughout the 50 cm of peat examined, the majority of TP was present in the form of total organic phosphorus. The pattern of TP_o content with depth matched that described for TP, with an enriched surface layer overlying peat with lower and more constant TP_o concentrations below 30 cm. TP_i content increased with depth from the surface to reach a maximum of approximately 400 mg kg^{-1} P between 10 and 15 cm, before decreasing towards the base of the core.

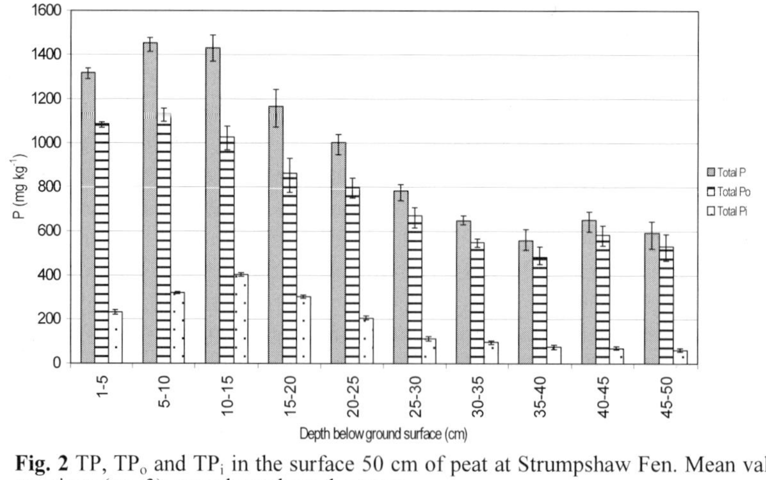

Fig. 2 TP, TP_o and TP_i in the surface 50 cm of peat at Strumpshaw Fen. Mean values are given (n = 3); error bars show the range.

Analysis of the peat from Strumpshaw Fen suggests that a large store of total phosphorus is contained within the surface 50 cm of sediment in this system. Comparison between the sediments at Strumpshaw Fen and other wetland and non-wetland deposits provides an indication of the degree of P-enrichment. The maximum total phosphorus content in the sediment at Strumpshaw Fen (>1400 mg kg^{-1}) lies towards the very top end of concentrations determined for organic deposits that have been affected by nutrient loading from anthropic activities such as agricultural runoff, industrial effluent or discharge from urban areas (e.g. Reddy *et al.*, 1998; Mayer *et al.*, 1999; Fisher & Reddy, 2001; Newman & Pietro, 2001; Soto-Jimenez *et al.*, 2003), and is certainly far in excess of the TP content of peat deposits unaffected by external nutrient loading, where concentrations below 500 mg kg^{-1} are common (e.g. White & Reddy, 1999; Craft & Casey, 2000; Fisher & Reddy, 2001; Wright *et al.*, 2001). As part of the flood plain system of the River Yare, the sediments at Strumpshaw Fen bear comparison with other non-wetland flood plain deposits. For example, in a study of 20 river systems in the UK, Walling *et al.* (2000) determined a mean TP content for flood plain sediments of 1247 mg kg^{-1} with a range of 417 to 2660 mg kg^{-1}. The most likely cause of the P-enrichment of the near-surface peat at Strumpshaw Fen is historical nutrient loading from the adjacent River Yare and Lackford Run. Monitoring of the River Yare indicates that during the period 1989–2002 the average MRP concentration was 0.84 mg l^{-1} P with a maximum concentration of 4.03 mg l^{-1} P (Environment Agency, unpublished data).

The release of P to solution following re-flooding of drained peat cores

The high concentration of P within the peat sediments at Strumpshaw Fen suggests a potential for release to solution, and thereby the maintenance of nutrient-enriched conditions within the wetland. Figure 3 shows the release of MRP to the surface water and pore water of a core following re-flooding with de-ionized water. The second core used in Experiment A produced very similar spatial and temporal patterns of MRP release to solution. Within six hours of re-flooding the core, the surface-water MRP concentration was greater than 0.1 mg l^{-1} P, and by the end of the first 24 hours of the incubation the MRP concentration was approximately 0.14 mg l^{-1} P. The release of MRP into the surface water continued throughout the 28 days (672 hours) of the experiment, resulting in a final concentration of 0.79 mg l^{-1} P. At sample depths of 17.5, 32.5 and 47.5 cm, pore-water concentrations of MRP remained near to or below the limit of detection during the first 24 hours of flooding. Samples from 2.5 and 10 cm depth contained approximately 0.08 and 0.05 mg l^{-1} P respectively, and these concentrations remained constant over the first 24 hours. Increases in pore-water MRP concentration in each sample began between 24 and 72 hours after flooding, and continued until approximately 360 hours, after which the samples tended to show less pronounced increases, or indeed decreases, in concentration. Pore-water MRP concentrations at the end of the period of flooding were highest in the sample from 17.5 cm where concentrations after 504 hours exceeded 2.3 mg l^{-1} P. Towards the peat surface, pore-water samples contained progressively lower concentrations of MRP. At 10 cm depth the average MRP concentration after 504 hours was 1.58 mg l^{-1} P, whilst at

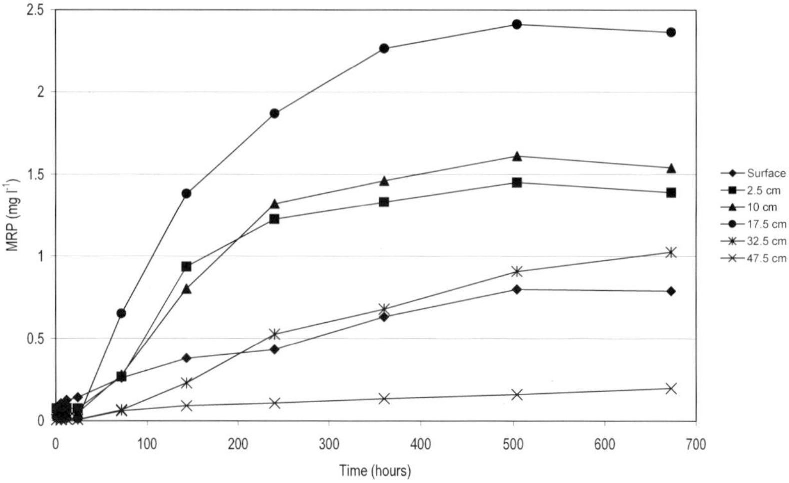

Fig. 3 Pore-water and surface-water MRP dynamics following re-flooding of peat cores with de-ionized water.

2.5 cm the average was 1.42 mg l^{-1} P. At depths below 17.5 cm there was also a progression to lower MRP concentrations: at 32.5 cm the average concentration was 0.97 mg l^{-1} P, and at 47.5 cm, 0.18 mg l^{-1} P.

The final surface-water and pore-water MRP concentrations in Fig. 3 exceed both the interim standard for eutrophic river conditions set by the UK Environment Agency (0.2 mg l^{-1} P), and the contemporary average concentrations in the River Yare and Lackford Run (Environment Agency, 2000). Given that the solution used to flood the cores in Experiment A was de-ionized water, these findings illustrate that the sediments at Strumpshaw Fen may act as a nutrient source to the wetland itself, even if external nutrient loadings to the site were to be reduced. More significantly, the scale of P release from the peat cores indicates that, given hydrological connection, the wetland system must be seen as a phosphorus source area, able to supply bioavailable P to receiving water bodies. This finding challenges the common conception of wetlands as nutrient sinks in the catchment, protecting the chemical and ecological quality of receiving waters.

The control of P release from riparian wetland sediments

In Experiments A and B the release of MRP to solution was strongly correlated with the release of ferrous iron (Fe^{2+}) ($r^2 \geq 0.96$, $P < 0.01$). Figure 4 shows the correlation of these two elements in all pore-water samples collected from both cores during Experiment B. The dominant P:Fe release ratio to pore water during the course of Experiment B equalled 0.463, and was consistent for all pore-water samples from both cores. The results of several lines of investigation cited in Gunnars & Blomqvist (1997) suggest that the predominant P:Fe surface complexation ratio of ferric hydroxide is 0.5. Given that the reductive-dissolution of Fe-P complexes should result in the

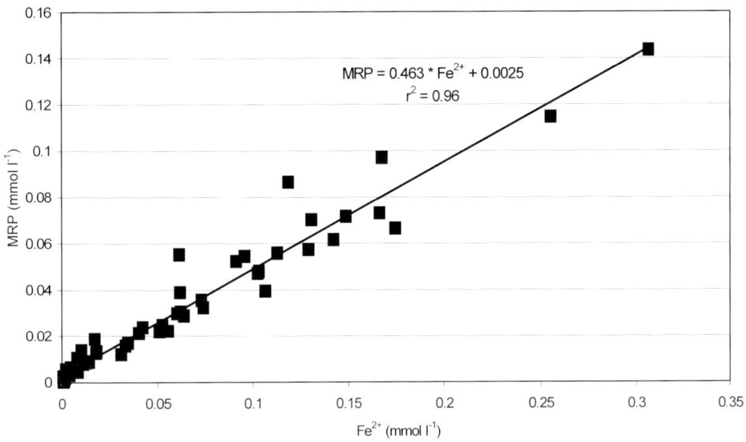

Fig. 4 The correlation of MRP and Fe^{2+} in all pore-water samples collected during Experiment B.

simultaneous, stoichiometric release of Fe and P (Gächter *et al.*, 1988), it is possible that MRP and Fe^{2+} were released into solution in the cores from Strumpshaw Fen during the reductive-dissolution of ferric hydroxide under anoxic conditions. Although the interaction between iron and phosphorus has been identified in other environments for some time (e.g. Lovley *et al.*, 1991; Gunnars & Blomqvist, 1997; Moore *et al.*, 1998; Gunnars *et al.*, 2002), very few studies have indicated a significant role for Fe-P in wetland sediments, often because of the low mineral content of many organic deposits (McLatchey & Reddy, 1998; Moore *et al.*, 1998). Yet, many riparian wetland systems have the potential to receive significant mineral loads, due to their hydrological connection with adjacent channel systems, which are subsequently incorporated within the peat matrix. In the sediments from Strumpshaw Fen the dynamics of iron, as driven by redox conditions, appear to control P release, rather than factors such as the mineralization of organic phosphorus. A dominant role for iron-bound phosphorus at Strumpshaw Fen is further supported by analysis of the changes to the sediment-P fractions as a result of core flooding. Compared to non-flooded cores, there was clear evidence that the iron-bound fraction in the sediment phosphorus pool was significantly depleted during the incubation of surface-inundated cores ($P < 0.01$), with the magnitude of the depletion corresponding well with the distribution of MRP in pore-water samples at the end of Experiment B.

CONCLUSIONS

This paper has shown that riparian wetland sediments may contain a large store of sediment-bound phosphorus, sequestered primarily from inputs of nutrient-enriched channel water. A proportion of this store of P is in a form that may be released into pore water and surface water following flooding. The release appears to be physicochemically controlled and involves the reductive-dissolution of Fe-P. The sediments at Strumpshaw Fen have the potential to act as phosphorus sources, establishing a nutrient-enriched status within the wetland even if external loads of P are reduced or removed entirely. Further, given hydrological connection, surface water *and* pore water from Strumpshaw Fen may deliver ecologically significant quantities of bioavailable phosphorus to receiving water bodies. Future research should focus on accurately quantifying the exchanges of water and associated solutes between wetlands and other catchment elements (e.g. channel networks, deeper ground water), and on assessing the fate of P following delivery to receiving water bodies.

Acknowledgements We are grateful to the Royal Society for the Protection of Birds and their reserve warden, Tim Strudwick, for allowing us to use Strumpshaw Fen as a research site. The research was supported by a studentship (to BWJS) from the University of Sheffield and private funds (AJB).

REFERENCES

Aerts, R., Verhoeven, J. T. A. & Whigham, D. F. (1999) Plant-mediated controls on nutrient cycling in temperate fens and bogs. *Ecology* **80**, 2170–2181.

Burt T. P., Pinay, G., Matheson, F. E., Haycock, N. E., Butturini, A., Clement, J. C., Danielescu, S., Dowrick, D. J., Hefting, M. M, Hillbricht-Ilkowska, A. & Maitre, V. (2002) Water table fluctuations in the riparian zone: comparative results from a pan-European experiment. *J. Hydrol.* **265**, 129–148.

Craft, C. B. & Casey, W. P. (2000) Sediment and nutrient accumulation in floodplain and depressional freshwater wetlands of Georgia, USA. *Wetlands* **20**, 323–332.

Environment Agency (2000) *Aquatic Eutrophication in England and Wales: A Management Strategy*. Environmental Issues Series. Environment Agency, Bristol, UK.

European Parliament and Council of the European Union (2000) Water Framework Directive 2000/60/EC. *J. European Communities.* Brussels, Belgium.

Fisher, M. M. & Reddy, K. R. (2001) Phosphorus flux from wetland soils affected by long-term nutrient loading. *J. Environ. Qual.* **30**, 261–271.

Gächter, R, Meyer, J. S. & Mares, A. (1988) Contribution of bacteria to release and fixation of phosphorus in lake sediments. *Limnol. Oceanogr.* **33**, 1542–1559.

Gunnars A. & Blomqvist, A. (1997) Phosphate exchange across the sediment-water interface when shifting from anoxic to oxic conditions – an experimental comparison of freshwater and brackish-marine systems. *Biogeochemistry* **37**, 203–226.

Gunnars, A., Blomqvist, S., Johansson, P. & Andersson, C. (2002) Formation of Fe(III) oxyhydroxides colloids in freshwater and brackish seawater, with incorporation of phosphate and calcium. *Geochim. Cosmochim. Acta* **66**, 745–758.

Lovley, D. R., Phillips, E. J. P. & Lonergan, D. J. (1991) Enzymatic versus non-enzymatic mechanisms for Fe(III) reduction in aquatic sediments. *Environ. Sci. Technol.* **25**, 1062–1067.

Mayer, T., Ptacek, C. & Zanini, L. (1999) Sediments as a source of nutrients to hypereutrophic marshes of Point Pelee, Ontario, Canada. *Water Res.* **33**, 1460–1470.

McLatchey, G. P. & Reddy, K. R. (1998) Regulation of organic matter decomposition and nutrient release in a wetland soil. *J. Environ. Qual.* **27**, 1268–1274.

Moore, P. A., Reddy, K. R. & Fisher, M. M. (1998) Phosphorus flux between sediment and overlying water in Lake Okeechobee, Florida: spatial and temporal variations. *J. Environ. Qual.* **27**, 1428–1439.

Newman, S. & Pietro, K. (2001) Phosphorus storage and release in response to flooding: implications for Everglades stormwater treatment areas. *Ecol. Engng* **18**, 23–38.

Novak, J. M., Stone, K. C., Szogi, A. A., Watts, D. W. & Johnson, M. H. (2004) Dissolved phosphorus retention and release from a coastal plain in-stream wetland. *J. Environ. Qual.* **33**, 394–401.

Owens, P. H. & Walling, D. E. (2002) The phosphorus content of fluvial sediment in rural and industrialized river basins. *Water Res.* **36**, 685–701.

Pant, H. K. & Reddy, K. R. (2003) Potential internal loading of phosphorus in a wetland constructed in agricultural land. *Water Res.* **37**, 965–972.

Pant, H. K, Reddy, K. R. & Spechler, R. M. (2002) Phosphorus retention in soils from a prospective constructed wetland site: environmental implications. *Soil Sci.* **167**, 607–615.

Prior, H. & Johnes, P. J. (2002) Regulation of surface water quality in a Cretaceous Chalk catchment, UK: an assessment of the relative importance of instream and wetland processes. *Sci. Total Environ.* **282–283**, 159–174.

Reddy, K. R., Wang, Y., DeBusk, W. F., Fisher, M. M. & Newman, S. (1998) Forms of soil phosphorus in selected hydrologic units of the Florida Everglades. *Soil Sci. Soc. Am. J.* **62**, 1134–1147.

Reddy, K. R., Kadlec, R. H., Flaig, E. & Gale, P. M. (1999) Phosphorus retention in streams and wetlands: a review. *Critical Reviews in Environmental Science and Technology* **29**, 83–146.

Soto-Jimenez, M. F., Paez-Osuna, F. & Bojorquez-Leyva, H. (2003) Nutrient cycling at the sediment-water interface and in sediments at Chiricahueto marsh: a subtropical ecosystem associated with agricultural land uses. *Water Res.* **37**, 719–728.

Venterink, H. O., Wiegman, F., Van der Lee, G. E. M. & Vermaat, J. E. (2003) Role of active floodplains for nutrient retention in the River Rhine. *J. Environ. Qual.* **32**, 1430–1435.

Walling, D. E., He, Q. & Blake, W. H. (2000) River flood plains as phosphorus sinks. In: *The Role of Erosion and Sediment Transport in Nutrient and Contaminant Transfer* (ed. by M. Stone), 211–218. IAHS Publ. 263. IAHS Press, Wallingford, UK.

White, J. R. & Reddy, K. R. (1999) Influence of nitrate and phosphorus loading on denitrifying enzyme activity in Everglades wetland soils. *Soil Sci. Soc. Am. J.* **63**, 1945–1954.

White, J. R., Reddy, K. R. & Moustafa, M. Z. (2004) Influence of hydrologic regime and vegetation on phosphorus retention in Everglades stormwater treatment area wetlands. *Hydrol. Processes* **18**, 343–355.

Wright, R. B., Lockaby, B. G. & Walbridge, M. R. (2001) Phosphorus availability in an artificially flooded southeastern floodplain forest soil. *Soil Sci. Soc. Am. J.* **65**, 1293–1302.

Dynamics and Biogeochemistry of River Corridors and Wetlands (Proceedings of symposium S4 held during the Seventh IAHS Scientific Assembly at Foz do Iguaçu, Brazil, April 2005). IAHS Publ. 294, 2005.

Modelling the fate of nitrate in a lowland catchment system

GUNTER WRIEDT[1], JORIS SPINDLER[1], HELMUT GEISTLINGER[2] & MICHAEL RODE[1]

1 *UFZ Environmental Research Centre Leipzig-Halle, Brueckstrasse 3a, D-39114 Magdeburg, Germany*
gunter.wriedt@ufz.de

2 *UFZ Environmental Research Centre Leipzig-Halle, Theodor-Lieser-Straße 4, D-06120 Halle, Germany*

Abstract Biogeochemical transformations and hydraulic processes in soil and groundwater can moderate nitrate loads in surface waters compared to nitrogen inputs to soil. A modelling approach was developed based on the loose coupling of a soil nitrogen model with a groundwater flow and a multi-species reactive groundwater-transport model. This approach combines transport time with availability of reaction partners and nitrate turnover. A case study was conducted in a small lowland catchment (Schaugraben, ~20 km^2) close to Osterburg in the north of Saxony-Anhalt, Germany, and demonstrated the spatial heterogeneity of nitrate loads and concentrations resulting from the spatial distribution of nitrate inputs, the arrangement of input areas to the channel network, and the consumption of reactive substances in the aquifer. An extension of this approach aims to describe turnover processes at the groundwater–surface water interface in order to close the gap between groundwater exfiltration and catchment outlet.

Key words groundwater; lowland catchment; modelling; nitrate; nitrogen; reactive transport; soil

INTRODUCTION

Diffuse nitrate pollution is a threat to ground and surface water quality. Chemical, biological and hydrological processes, and transport pathways can diminish nitrate concentration in water bodies when compared to landscape nitrogen (N) inputs. Such processes include denitrification of nitrate in soil or groundwater, mixing of water bodies, retardation due to travel times of decades, and the spatial arrangement of flow paths, reactive substances and nitrate sources. These processes are difficult to observe in the field and therefore little is known about their spatial and temporal interaction. However, models are useful for investigating these interactions based on prior knowledge or experimental findings. Recent development of reactive groundwater transport models such as RT3D (Clement, 1997) allows the implementation of arbitrary reaction systems, and facilitates process-oriented studies of nutrient transport in lowland catchment systems. Through a combination of selected models we simulated nitrate transport in lowland catchments and provided improved understanding of N transport in soils and groundwater bodies. With this approach we investigated interactions between nitrate transport and chemical and catchment hydrological characteristics. In this paper, we provide an outline of the modelling approach and present a case study of the model application to a small lowland catchment, Schaugraben, Germany. Based on the results of this work, we discuss other potential applications of the modelling approach as well as further extensions, as not all aspects of the nitrogen transport system, such as hyporheic zone and in-stream transformations, are currently implemented.

MATERIAL AND METHODS

Outline of the integrative modelling concept

The modelling approach is based on a combination of submodels that collectively simulate water flow, N transport and N turnover in soil and groundwater. The models are loosely coupled, with results from one sub-model used as input for another sub-model. Soil processes were simulated using mRISK-N (Wriedt, 2004), which is a combination of the SIMPEL soil water balance model by Hörmann (1998) and the soil N model RISK-N (Gusman & Marino, 1999). Groundwater flow and solute transport are simulated using MODFLOW (McDonald & Harbough, 1998) and RT3D (Clement, 1997). The soil N sub-model considers the major N transformations including mineralization, nitrification and denitrification in soil systems, and calculates leaching of nitrate to groundwater. To include nitrate transformations in the groundwater, a special reaction module was developed for the RT3D model (Wriedt, 2004). This reaction module includes the decay of organic matter by oxygen, nitrate and sulfate as well as the oxidation of pyrite by oxygen and nitrate. This approach considers denitrification as the main process as well as a general geochemical framework based on kinetic reactions. With this set of tools it is possible to link chemical and hydraulic sediment characteristics to nitrate turnover. Special tools have been developed to support grid based distributed soil simulation, data transfer between the models and analysis of model results. A general diagram of the modelling approach is shown in Fig. 1.

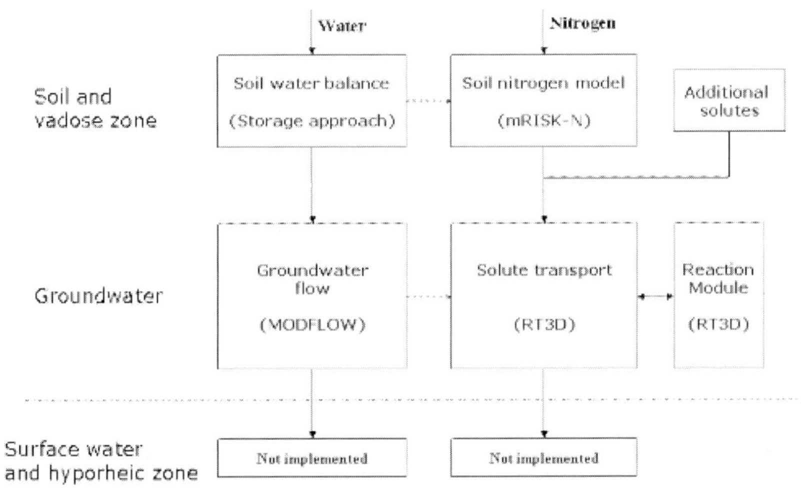

Fig. 1 Modelling approach.

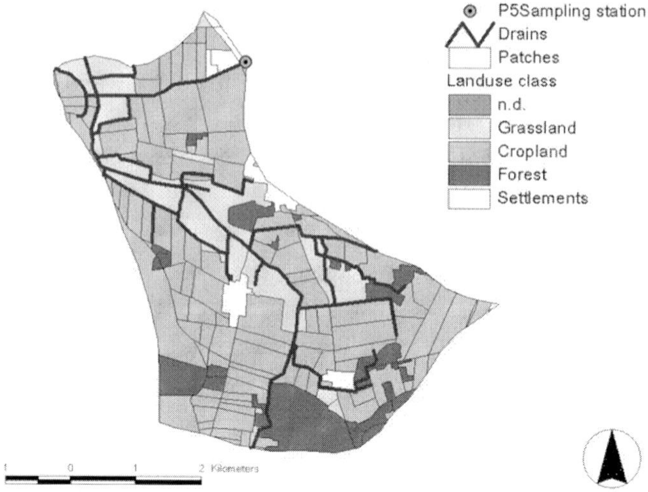

Fig. 2 Map of the study area, the Schaugraben catchment.

Study area

The case study was based on data from the Schaugraben catchment close to Osterburg (Altmark) in the north of Saxony-Anhalt, Germany. The Schaugraben catchment covers an area of 20 km^2 and comprises glacial till and glacio-fluvial deposits. The dominant soils are gleyic and histic soils in the lowlands, and cambisols and luvisols in the higher parts of the catchment. Land use is mainly agricultural and is artificially drained using a system of open drain channels. Land use and land management practice information has been collected since 1990, and catchment discharge and contaminant loads have been collected at a central gauging station P5 (Fig. 2) since 1997. A variety of research has been undertaken, including investigations into nitrate transport in soils and groundwater (Meissner, 2000). A map of the study area is given in Fig. 2.

Simulation of water and nitrogen fluxes at the catchment scale

Catchment scale simulations aim to investigate the potential of the modelling approach to imitate nitrate transport, and to study the effect of altering the spatial distribution of nitrate input on exfiltration to surface waters. Climate records, soil parameters (field capacity, hydraulic conductivity) and land use data provided input to a soil model. Figure 3 illustrates the average fertilizer input in the catchment area. Nitrate balance data for different land use classes are given in Table 1. Borehole profiles were used to construct a hydrogeological model of substrate characteristics to a depth of 40–50 m, providing a basis for the regional three-dimensional groundwater flow. The soil water and groundwater flow simulation was first set up for an extended study area, which

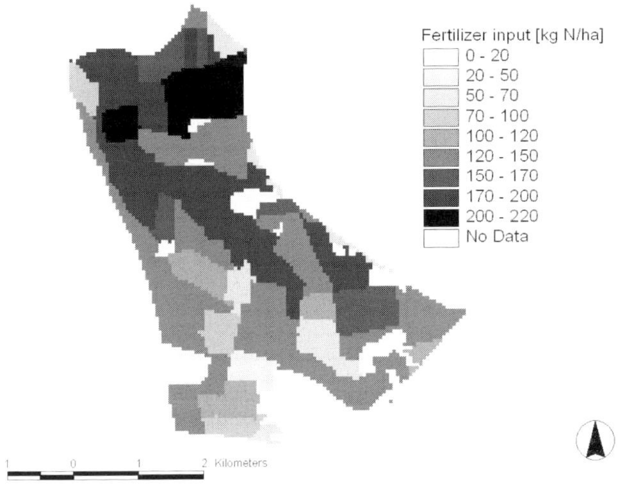

Fig. 3 Fertilizer input (kg N ha^{-1} year^{-1}) in the study catchment based upon a survey of management data from 1990 till 2003.

Table 1 Average nitrate inputs and simulated outputs for land use classes in the study catchment for observation period 1990–2003.

Land use	Area (ha)	Simulation input data: Atmospheric deposition[a] (kg N/ha)	Mineral fertilizer[a] (kg N/ha)	Organic fertilizer[a] (kg N/ha)	Gross plant uptake[b] (kg N/ha)	Simulation results: Groundwater recharge (mm)	Nitrate leaching (kg N/ha)
Pasture	345 (18%)	60	176	8	210	64±14	80±5
Arable land	1258 (66%)	60	108±25	35±24	165±39	89±12	160±11
Forest	258 (13%)	60	0	0	173	49±14	30±2
Settlement	59 (3%)	60	60	0		39±5	30±2

[a] Data given by mean and standard deviation result from observed data. Where the mean value is given only, data are homogeneous for land use classes and based upon observations or literature review.

[b] Gross plant uptake equals loss by harvest and N storage in remaining biomass, based upon data taken from the CANDY model data base (Franko, 1995).

extends the study catchment to the surrounding rivers and channels in order to get well defined model boundaries. This simulation served as a base for delineating the catchment boundaries. In a second step a refined model of the study catchment was set up and nitrate transport was subsequently added, simulating soil nitrogen dynamics and groundwater nitrogen dynamics separately using soil simulation results as input for the groundwater simulation. The model was validated from observed fluxes and groundwater levels, lysimeter data and observed solute concentrations in ground and surface waters.

The influence of three different N input scenarios on nitrate leaching was examined. The scenarios included distributed N input, uniform N input based on average rates, and distributed N input with a buffer zone of 100 m width along the channel system, where no nitrate leaching occurs. The uniform input scenario is based on the same data as the distributed scenario, but N input to groundwater was averaged over the entire study area. All scenarios were estimated using reactive and conservative transport conditions, assuming constant land use (see Figs 2 and 3) over a period of up to 200 years. The groundwater reaction module was parameterized based on data in the literature (Böttcher *et al.*, 1990; Hunter *et al.*, 1998; Parkhurst & Appelo, 1999; Kameih & Ohmoto, 2000; Mayer, 2000; MacQuarrie & Sudicky, 2001). The distribution of pyrite and sedimentary organic matter in the aquifer was considered homogeneous, except below gleyic and histic soils where organic matter contents were higher. We evaluated the distribution of nitrate in groundwater at different points in time, the spatial differentiation of flow and nitrate loads in the groundwater exfiltrating into the channel system, and the temporal development of average nitrate concentrations in the channel system resulting from groundwater exfiltration. Groundwater inputs were considered constant during the simulation period.

RESULTS

Under conditions of distributed N input, the distribution of nitrate concentrations in recharge water leaching from the soil (Fig. 4) reflects the different levels of fertilization of single fields. In areas close to the channel system, the presence of gleyic and histic soils rich in organic matter leads to reduced nitrate leaching from enhanced denitrification. Nitrate concentrations are distinctly higher further away from the channels presumably due to lower denitrification and/or higher fertilizer inputs.

Nitrate concentrations in the groundwater exfiltrating into the channel system show a distinct spatial heterogeneity (Fig. 5). Channel segments receiving highly polluted groundwater are closely related to agricultural areas with high nitrate inputs, located upstream of the groundwater flow. Non-polluted segments correspond with the areas of intense denitrification under gleyic and histic soils.

Nitrate concentrations in the exfiltrating groundwater follow an S-shaped breakthrough curve under conservative transport conditions (Fig. 6). After a simulation period of 200 years, the theoretical maximum of 70 mg l^{-1} nitrate corresponding to the average input concentration was not reached.

Under uniform N input the breakthrough curve is determined by travel times only, whereas in the distributed leaching scenario the spatial differentiation leads to deviations from this ideal curve. Including a 100-m buffer zone in the distributed N input scenario leads to a reduction of the total input which is reflected in a reduced maximum of the nitrate breakthrough curve. A second effect is retardation of nitrate breakthrough, as areas close to the drain system no longer contribute to the nitrate loads of the channel (Fig. 6).

Under reactive transport conditions, model behaviour is principally similar (Fig. 7), however, final concentrations after a 200-year simulation are much lower, reaching 17 mg l^{-1} nitrate under spatially distributed N input. Additionally the breakthrough curves do not approach a maximum concentration and it can be expected that nitrate concentrations will increase slowly but constantly, as long as reactive substances are present within the aquifer.

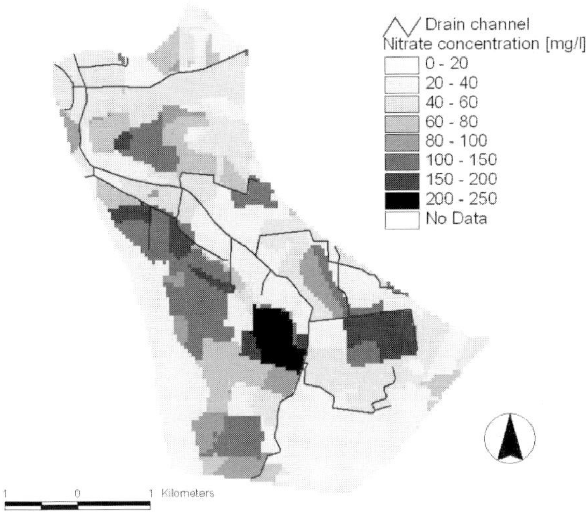

Fig. 4 Nitrate concentrations in the recharge water leaching from the soil.

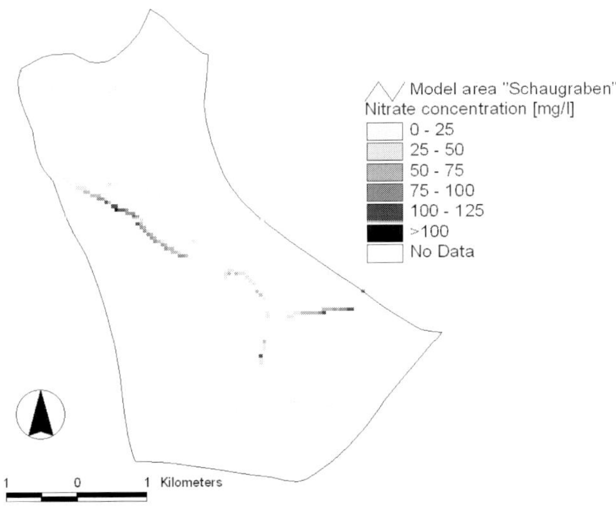

Fig. 5 Nitrate concentrations in the exfiltrating groundwater after a simulation period of 200 years under reactive transport conditions.

Figure 8 illustrates the balance of groundwater inputs and outputs of N as cumulative and daily loads. Under conservative transport, only 50% of the total (cumulative) input left the system as output to surface waters although output loads nearly balance input loads, due to storage within the aquifer. Under reactive transport conditions only 10% of the input has left the system at the end of the simulation period, due to the combined effect of aquifer storage and denitrification.

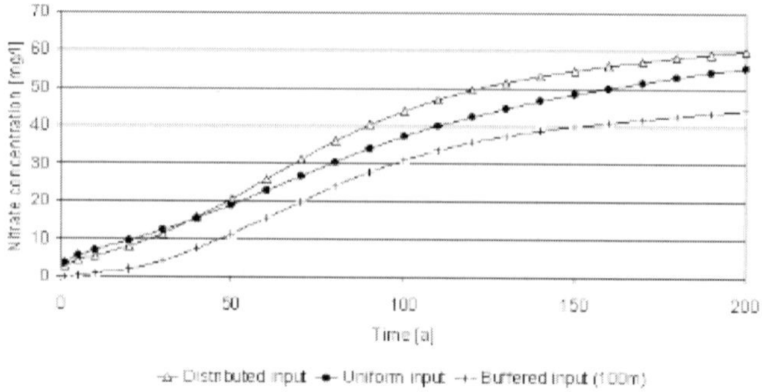

Fig. 6 Development of average nitrate concentrations in exfiltrating groundwater in the conservative transport simulations.

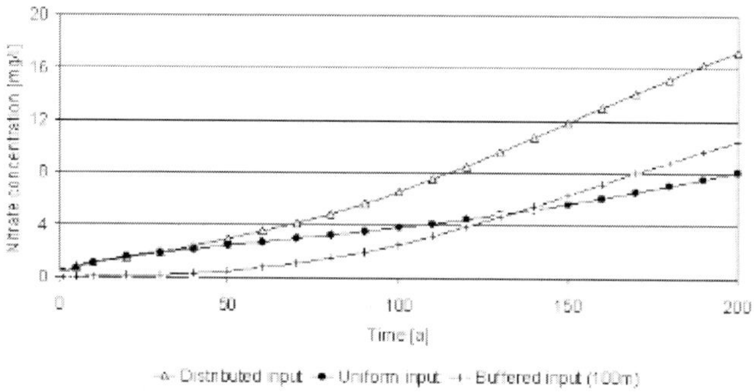

Fig. 7 Development of average nitrate concentrations in exfiltrating groundwater in the reactive transport simulations.

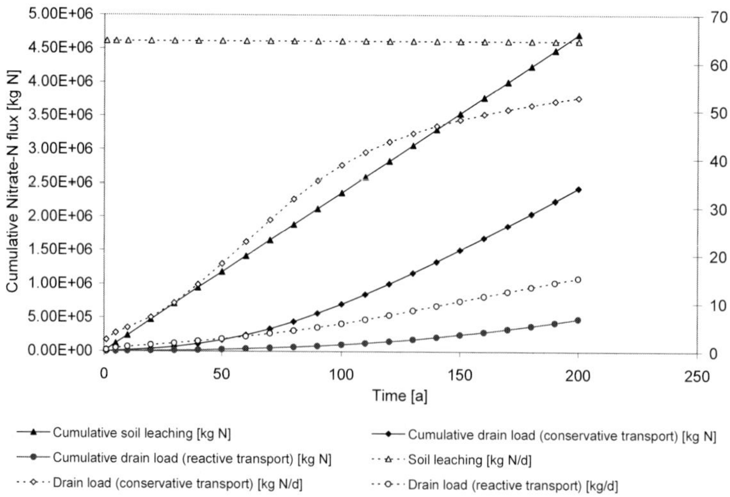

Fig. 8 Soil leaching of nitrate-N and output into the drain system as cumulative and daily loads under conservative and reactive transport conditions.

DISCUSSION

The modelling approach presented here allows simulation of various processes related to nitrate transport in lowland catchments, such as: (a) the distribution and turnover of solutes and immobile components in space and time, (b) the development of solute concentrations in the exfiltrating groundwater as a measure of groundwater contribution to the pollution of surface waters, and (c) the spatial distribution of groundwater exfiltration rates and solute concentrations contributed to the channel system. The spatially distributed approach and the consideration of reactive groundwater transport facilitate the investigation of transport and transformation interactions, and potential management opportunities. The validity of model results for a specific area, however, strongly depends on availability and uncertainty of input data, model parameters, and validation data, which are not further discussed in this paper. Despite these limitations, the approach allows identification of the main sources and sinks of nitrate pollution, allows water protection measures (land-use change and management) to be more targeted, supports design of further experimental studies and can be checked against localized field observations.

Further model modifications will be necessary to more fully investigate physical and chemical interactions at the catchment scale and the temporal dynamics of the processes. Observed nitrate concentrations in the Schaugraben channel range between 0 and 140 mg l^{-1} and follow an annual cycle with an average concentration of 32 mg l^{-1}. Field observations indicate that rising groundwater levels in winter increase the connectedness of contributing areas to the channel network, increasing nitrate pollution. In the summer, these areas are disconnected as the channels fall dry. Thus future studies will include time variant simulations in order to study the effect of intra-annual water level and nitrate variations. Further modifications to the modelling approach will be required to include transformations of nitrogen compounds in the hyporheic zone and open water bodies, atmospheric pathways and plant uptake and storage. The modular structure of the current approach allows these aspects to be incorporated simply.

Acknowledgements We gratefully acknowledge funding of this project by the German Research Foundation DFG.

REFERENCES

Böttcher, J., Strebel, O. & Duynisveld, W. H. M. (1989) Kinetik und Modellierung gekoppelter Stoffumsetzungen im Grundwasser eines Lockergesteins-Aquifers. *Geologisches Jahrbuch Reihe C* **51**, 3–40.

Clement, T. P. (1997) *RT3D – A Modular Computer Code for Simulating Reactive Multi-Species Transport in 3-Dimensional Groundwater Aquifers*. PNNL-11720. Pacific Northwest National Laboratory, Richland, Washington, USA.

Franko, U., Oehlschlägel, B. & Schenk, S. (1995) Modellierung von Bodenprozessen in Agrarlandschaften zur Untersuchung der Auswirkungen möglicher Klimaveränderungen. *UFZ-Bericht 1995(3).*

Gusman, A. J. & Marino, M. A. (1999) Analytical modeling of nitrogen dynamics in soils and groundwater. *J. Irrig. Drain. Engng* **11/12**, 330–337.

Hörmann, G. (1998) Simpel – Speichermodelle zum Bodenwasserhaushalt. Dokumentation zu Simpel 2, http://www.pz-oekosys.uni-kiel.de/~schorsch/simpel/index.htm

Hunter, K. S., Wang, Y. & Van Capellen, P. (1998) Kinetic modelling of microbially driven redox chemistry of subsurface environments: coupling transport, microbial metabolism and geochemsitry. *J. Hydrol.* **209**, 53–80.

Kamei, G. & Ohmoto, H. (2000) The kinetics of reactions between pyrite and O_2-bearing water revealed from *in-situ* monitoring of DO, Eh and pH in a closed system. *Geochimica et Cosmochimica Acta* **64**(15), 2585–2601.

MacQuarrie, K. T. B. & Sudicky, E. A. (2001) Multicomponent simulation of wastewater-derived nitrogen and carbon in shallow unconfined aquifers I. *J. Contam. Hydrol.* **47**, 53–84.

Mayer, K. U. (2000) MIN3P V1.1 User Guide. Unpublished.

McDonald, M. G. & Harbaugh, A. W. (1998) A modular three-dimensional finite-difference ground-water flow model. *US Geol. Survey Open File Report 83-875.*

Meissner (2000) Quantifizierung von diffusen Stoffausträgen aus der landwirtschaftlichen Flächennutzung in den Einzugsgebieten von Oka und Elbe. *UFZ-Bericht 2000(13).*

Parkhurst, D. L. & Appelo, C. A. J. (1999) User's guide to PHEEQC (Version 2) – A computer program for speciation, batch reaction, one-dimensional transport, and inverse geochemical calculations. *Water-Resources Investigations Report 99-4259*, Denver, USA.

Wriedt, G. (2004) Modelling of nitrogen transport and turnover during soil and groundwater passage in a small lowland catchment of Northern Germany. Dissertation, Potsdam University, Potsdam, Germany. http://pub.ub.uni-potsdam.de/2004meta/0030/door.htm

Oxbow lakes as biogeochemical filters for nutrient outflow from agricultural areas

KATARZYNA GLIŃSKA-LEWCZUK
Department of Land Reclamation and Environmental Management, University of Warmia and Mazury in Olsztyn, Plac Lodzki 2, Poland
kaga@uwm.edu.pl

Abstract The investigation of oxbow lakes conducted in the middle section of the River Łyna valley, which is representative of the post-glacial landscape in northeastern Poland, showed a susceptibility to degradation with regards to hydrological connectivity and external sources of supply. When the connection to a river is maintained, oxbow lakes prolong their existence, thanks to multiple exchange of matter. When oxbow lakes are disconnected, their degradation is much faster and they undergo advanced hypertrophy, overgrowth and eventually disappearance. The process is accelerated when an oxbow lake becomes a trap for nutrient effluents from reclamation systems that dewater highly-managed arable lands or grasslands. In agricultural river valley landscapes, oxbow lakes may function as very effective biogeochemical filters, contributing to sediment trapping, nutrient removal and river pollution prevention. However, the benefits are unilateral: ecological equilibrium in the oxbow lake ecosystem is easily lost.

Key words biogeochemical filter; connectivity; flood plain; nutrients; oxbow lake

INTRODUCTION

Oxbow lakes are an integral element of meandering rivers valleys. They often exchange water as well as dissolved and particulate matter through runoff and groundwater discharge phenomena, similar to other components of the river landscape. During the passage through the oxbow lakes, these substances are involved in complex biogeochemical processes, which may lead to chemical transformations with important consequences for the functioning of the ecosystems (Hein *et al.*, 2004; Tockner *et al.*, 1999).

An essential role in the functioning of aquatic systems within or between elements of river flood plains is played by hydrological connectivity (Amoros *et al.*, 2000; Amoros & Bornette, 2002). Ward *et al.* (2002) distinguished a diversity of lotic, semi-lotic and lentic water bodies occurring within river corridors whose permanence and succession relies on connectivity to the river.

A factor complicating the interactions between river and oxbow lakes is that the landscape has often been modified by man to improve drainage or prevent flooding. During the 20th century major rivers were intensively regulated for different purposes, including land reclamation and flood-control projects. In many cases, when the river straightening followed land reclamation, newly formed cut-offs, not the river itself, became direct recipients for the nutrient-rich water from reclaimed areas. Oxbow lakes, particularly those disconnected to the river, become traps for nutrient effluents from reclamation systems dewatering highly managed arable lands or grasslands.

The excess of nutrients influences habitat conditions of oxbow lakes and accelerates their natural susceptibility to degradation. This has become a major concern for both nature conservation groups and landowners due to their ecological values (Glińska-Lewczuk, 2004c). Awareness of the direct export of water from reclaimed areas into oxbow lakes seems to be a controversial solution for river protection against pollution of agricultural origin. Nevertheless, even though the river may be protected only to some extent, it eventually leads to the oxbow lake extermination.

This paper describes an approach to studying the biogeochemical interactions among oxbow lakes, ditch and river water in the middle section of the River Łyna in northeastern Poland. The objectives of the study were to compare water quality parameters of three types of oxbow lakes due to their hydrological connectivity, and also provide information on the accumulation of dissolved substances in relation to the external input of nutrients from reclaimed agriculturally used areas.

MATERIAL AND METHODS

Study area

The present study was conducted in the free-flowing section of the River Łyna in northeastern Poland (Fig. 1). At Smolajny village, the River Łyna drains a 2290-km^2 area and its average discharge is 14 m^3 s^{-1}, ranging from 7 m^3 s^{-1} to 35 m^3 s^{-1}. Floods have a recurrence period of 1.4 years, calculated over a 40-year cycle. They take place in the early spring months due to the snowmelt. Within the period of observation from January 2002 to August 2004, only two flood events were recorded: the first one lasted approx. 2 months and finished in mid-March 2002; the second one lasted 20 days and occurred in mid-February 2004. The peak discharge during the first flood was 31 m^3 s^{-1}, whereas during the second one was 27 m^3 s^{-1}.

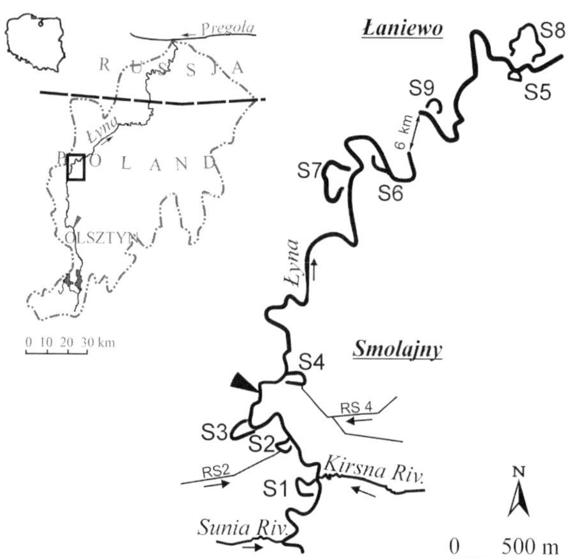

Fig. 1 Location of the study area.

The river valley is at an altitude of 70 m a.m.s.l. The flat flood plain, lying approx. 1.4 km on both sides of the river, constitutes an area prone to flooding on a 15-km stretch of the river, between the villages of Smolajny and Łaniewo. The river landscape of the River Łyna has changed over the last century. Before 1940 the river valley was reclaimed. At the end of the 1970s, the 7-km long section near Smolajny was straightened to prevent agricultural areas from flooding.

The oxbow lakes located in the River Łyna valley are elongated and have single-loops inherited from the former river channel. Their lengths are 200–400 m and the average sinuosity is 3.0. Their water table areas are relatively small (approx. 1 ha) and depths are below 2.5 m (when referred to the water level at 100 cm on gauge). The water resources of oxbow lakes are heavily dependent on the river water table fluctuations, which amounted to 201 cm in the period of observation.

For the purpose of this study, nine study sites were selected from all 60 oxbow lakes located nearby Smolajny village. Their morphological properties are shown in Table 1. The chosen sites were divided into three particular groups with respect to the degree of hydrological connectivity: open, semi-open and closed. The features of each group are summarized in Table 2.

Two of the studied oxbow lakes (S2 and S4) are direct recipients of ditch water. RS2 is the ditch which dewaters a 52-ha area of arable land towards S2. The outflow by ditch RS4 is from 249 ha of arable land in the upper part and grassland in the lower part (flood plain) of the ditch basin. The outflows by both ditches were relatively stable

Table 1 Morphological properties of the studied oxbow lakes.

Parameter	Unit	Oxbow lake:								
		S1	S2	S3	S4	S5	S6	S7	S8	S9
Water table area, A	ha	1.58	0.92	1.07	0.91	0.31	0.38	0.98	1.58	0.25
Storage capacity, C	10^3 m^3	16.5	10.7	22.8	9.9	5.7	2.2	1.8	7.2	3.5
Maximal length, L	m	920	503	473	493	256	247	635	1050	183
Maximal width, W	m	26.4	24.7	28.0	23.2	16.3	22.8	19.8	36.8	28.3
Elongation index, L/W	–	34.8	20.4	16.9	21.3	15.7	10.8	32.1	28.5	6.5
Maximal depth, D_{max}	m	3.1	2.6	4.2	2.4	2.3	2.2	1.1	1.4	1.4

Table 2 Hydrological connectivity: summary of characteristics and corresponding functional properties of oxbow lakes in the River Łyna valley.

Parameter	Type of oxbow lake:		
	Open	Semi-open	Closed
Hydrological connectivity	Lotic	Semi-lotic	Lentic
	Upstream and downstream connections	Downstream connection	Disconnected
Retention time	Short (0.5–7 days)	Long (12–40 days)	Very long (∞)
Transparency	Low	Moderate	High
Trophic state	Mezotrophy, Eutrophy	Eutrophy, Hypertrophy	Hypertrophy
Nutrient dynamics	Open system cycling	Open system cycling	Closed system cycling
Nutrient source	Allogenic	Autogenic/allogenic prevail when ditch and river supply	Autogenic
Oxbow lake no.	S3, S5	S1, S2, S4, S6	S7, S8, S9

and permanent. The mean water discharge in ditch RS4 in the period of investigation was 4.95 l s^{-1} (3.15–7.71 l s^{-1}), whereas RS2 was relatively small and amounted 1.10 l s^{-1} (0.52–1.92 l s^{-1}). The annual outflow volume of RS2 was three times higher and RS4 12 times higher than the storage capacities of the oxbow lakes they supply.

Hydrological measurements

Water levels in the River Łyna were recorded at Smolajny by a reference D-Diver (Ejkelkamp, The Netherlands) employed in the river 100 m upstream from the oxbow lake S4. The fluctuations in the water level in oxbow lakes were assumed to be identical to those in the river. Discharges of ditch water were measured simultaneously with the time of sampling using an electromagnetic flow velocity sensor (Valeport, UK) and calculated with the velocity–area method. The morphological parameters of oxbow lakes were based on bathymetry measurements described in detail by Glińska-Lewczuk *et al.* (2004a).

Water sampling and analytical procedure

Water sampling was carried out on a monthly basis from nine sampling sites located on oxbow lakes (S1–S9), two ditches (RS2, RS4), and one river. Water temperature (T, °C), electrical conductivity (EC, µS cm^{-1}), dissolved oxygen (DO, mg l^{-1}) and pH were recorded using portable meters (WTW). The laboratory analyses of water included mineral forms of nitrogen (TIN): N-NO_2, N-NO_3 and N-NH_4 as well as TP and P-PO_4 total dissolved matter (TDS) and mineral matter (MM). Organic matter (OM) concentration was calculated as subtraction of TDS–MM. Laboratory analyses followed the standards described by Hermanowicz *et al.* (1999). Fluxes of dissolved determinants in ditch water were calculated as the product of the concentration and corresponding discharge using the method presented by Webb *et al.* (2000). Accumulation of a given substance in oxbow lake water was calculated on the basis of its concentration and water volume at the time of sampling. The total annual load from a ditch basin was converted to the accumulation in an oxbow lake and expressed in g per 1 m^2 of water table area.

RESULTS

Physicochemical properties of water

The oxbow lakes in the River Łyna valley have water with near neutral pH values, and are rich in dissolved solids, but have more variable nutrients concentrations than the river itself. Table 3 summarizes the mean values of the physicochemical parameters of the studied oxbow lakes, river and ditch water. Although all types of water bodies showed an average degree of mineralization, the oxbow lakes had the lowest values when expressed in both electrical conductivity and mineral matter concentration. The comparison of groups of means shows that oxbow lake water is significantly different from river or ditch water with regards to P-PO_4, N-NO_2 and N-NO_3.

Table 3 An overview of the physical and chemical properties of river, oxbow lakes and ditch water (mean ± SD) within the River Łyna valley near Smolajny in the period January 2002–August 2004.

Object	Temp	pH	EC	O_2	TDS	MM	OM	$P\text{-}PO_4$	TP	$N\text{-}NO_2$	$N\text{-}NO_3$	$N\text{-}NH_4$
River Łyna at Smolajny	12.41 ± 6.36	7.44 ± 0.23	$396^{I/II}$ ± 34.26	7.85 ± 1.68	292^{I} ± 28.03	212^{I} ± 34.81	80.2^{I} ± 3643	$0.61^{I/II}$ ± 0.22	1.23 ± 0.89	0.030^{II} ± 0.02	0.81^{II} ± 0.37	0.62^{II} ± 0.50
Reclamation ditch	9.27 ± 4.49	9.91 ± 16.36	499^{II} ± 213.3	6.58 ± 3.01	357^{II} ± 50.50	250^{II} ± 43.86	104.0^{II} ± 43.21	0.76^{II} ± 0.60	1.36 ± 0.70	0.019^{II} ± 0.02	1.18^{III} ± 0.81	0.20^{I} ± 0.23
Oxbow lakes (all objects)	12.09 ± 7.27	7.44 ± 0.44	401^{I} ± 67.32	7.74 ± 3.42	305^{I} ± 57.81	218^{I} ± 47.98	$91.7^{I/II}$ ± 50.20	0.50^{I} ± 0.46	1.17 ± 0.91	0.018^{I} ± 0.01	0.47^{I} ± 0.45	$0.47^{I/II}$ ± 0.60
Open	9.88 ± 7.23	7.53^{A} ± 0.38	367^{B} ± 30.36	7.89^{B} ± 1.69	279 ± 12.01	211 ± 12.29	87.2 ± 64.13	0.43^{A} ± 0.20	1.04^{A} ± 0.51	0.028^{B} ± 0.01	0.87^{B} ± 0.48	0.62 ± 0.63
Semi-open	12.10 ± 7.52	7.46^{A} ± 0.47	413^{B} ± 64.46	7.96^{B} ± 3.71	311 ± 63.53	221 ± 50.71	94.5 ± 51.94	0.40^{A} ± 0.29	1.05^{A} ± 0.75	0.015^{AB} ± 0.01	0.38^{AB} ± 0.43	0.43 ± 0.60
Closed	12.88 ± 7.87	6.75^{B} ± 0.33	247^{A} ± 94.79	3.00^{A} ± 2.62	287 ± 64.13	197 ± 70.85	90.0 ± 37.76	1.98^{B} ± 1.066	3.32^{B} ± 1.48	0.004^{A} ± 0.001	0.12^{A} ± 0.034	0.40 ± 0.18
Supplied by ditch	12.42 ± 7.46	7.29 ± 0.50	419 ± 54.45	8.47 ± 4.20	321^{a} ± 37.29	226 ± 37.81	96.4^{a} ± 40.54	0.53 ± 0.29	1.00 ± 0.47	0.018 ± 0.01	0.54^{a} ± 0.54	0.31 ± 0.47
No ditch supply	12.82 ± 7.62	7.46 ± 0.42	409 ± 53.15	7.88 ± 3.42	290^{b} ± 34.19	224 ± 50.11	75.6^{b} ± 20.84	0.43 ± 0.27	1.13 ± 0.81	0.008 ± 0.01	0.30^{b} ± 0.29	0.42 ± 0.52

*Groups of means significantly different in Duncan test at $P \leq 0.05$: $^{I, II, III}$, river, ditch, oxbow lakes; $^{A, B}$, open, semi-open and closed oxbow lakes. Groups of means significantly different in *test t* at $P \leq 0.05$: $^{a, b}$, concern oxbow lakes with and without ditch supply.

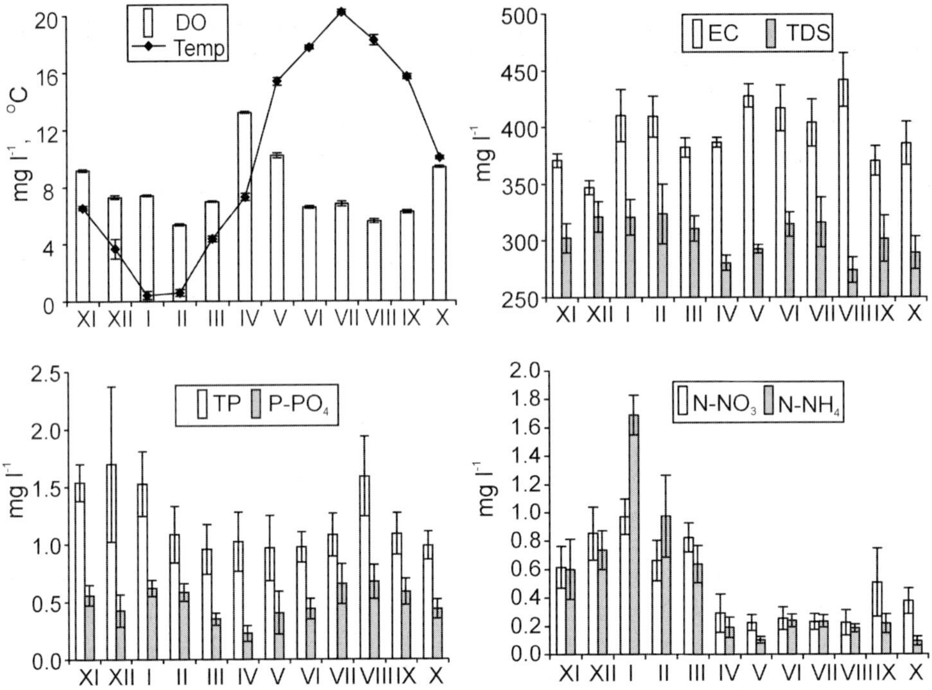

Fig. 2 Monthly changes of selected physicochemical properties (mean and standard error of mean) of oxbow lake water.

Due to the intensive changes in oxygen content in oxbow lakes, all determinants connected to biochemical processes exhibited distinct fluctuations over time and demonstrated a wide range of values (Fig. 2). Water temperature showed minimum values of >0.3°C in January and February and maximum values in July >20°C. Two seasons of oxygen decrease (5.0–7.0 mg l^{-1}) and, in the case of disconnected oxbow lakes, deficiency may be distinguished: the first one during snow and ice cover (December–March) and the second one during the warm, dry summer months (June–September). The concentration of dissolved oxygen increased rapidly just after ice cover melting in early spring (April). This "deep breath" caused a short-term over-saturation achieving 13.0 mg l^{-1}. During autumn, the second period of the oxygen increase was noted but was significantly lower in comparison to the spring period (9.0 mg l^{-1}).

TP concentrations in the oxbow lakes amounted on average to 1.17 mg l^{-1}, however the maximal values achieved were 5.3 mg l^{-1}. The data set showed a marked pattern in the seasonality of P concentrations. During summer, TP values were elevated due to internal P cycling that may have occurred and been supported by high summer water temperatures and pH (e.g. >pH 8). All of the sites had a commonly observed seasonal pattern of reduced P-PO_4 concentrations in spring (0.3 mg l^{-1}) with high levels throughout the summer (0.6 mg l^{-1}), and a subsequent decrease in autumn (0.3 mg l^{-1}) and higher again in winter (January–February). Although, TP and P-PO_4 concentrations were highly variable over the year at all sites, the highest range in values occurred at the highly enriched sites (lentic oxbow lakes).

In contrast to the varied seasonal distributions of TP and $P\text{-}PO_4$, $N\text{-}NO_3$ showed a remarkably consistent seasonality in all oxbow lakes. The concentrations of $N\text{-}NO_3$ were moderate (mean: 0.47 mg l^{-1}). Except for the sites S2, S4 which had high mean concentrations greater than 2 mg l^{-1}, most likely caused by agricultural runoff, the majority of the sites experienced late winter/early spring maxima (0.6–0.8 mg l^{-1}), followed by a period of low concentrations through the summer and autumn (0.2–0.3 mg l^{-1}). This seasonal pattern of events is usually attributed to increased rates of $N\text{-}NO_3$ assimilation by algae throughout the spring and summer growing season. During winter, $N\text{-}NO_3$ concentrations increase, as uptake by phytoplankton and denitrifying bacteria is minimal.

Effect of connectivity on chemical composition of oxbow lakes water

Three groups of oxbow lakes (open, semi-open, closed ones) showed significant differences in chemical parameters of water. Lotic habitats had parameters similar to river water, in particular to EC (Table 3). Very short-term water retention in lotic habitats is not favourable for sedimentation, thus mineral particles are easily moved. This is contrary to disconnected water bodies, where retention time is incomparably longer and closed system cycling occurs.

Oxbow lakes disconnected from the river showed significantly lower values of $N\text{-}NO_3$ (0.12 mg l^{-1}) and $N\text{-}NO_2$ (0.004 mg l^{-1}) in comparison to those connected to the river: 0.87, 0.028, respectively. On the contrary, under oxygen deficiency conditions, disconnected oxbow lakes maintained high levels of TP, on average 3.32 mg l^{-1}. An analysis of seasonal changes of TP in lentic oxbow lakes showed that significant amounts of P were released to water during spring. The connection with the river allows the other two types of oxbow lakes to maintain stable TP concentrations, however at a high level of approx. 1.0–1.5 mg l^{-1}.

External sources of oxbow lakes supply

River water table fluctuations and alluvial ground water movements determine changes in chemical budget of oxbow lakes. The rise in water levels from the lowest stages to bank over-flooding is favourable for nutrient and other chemical element exchanges and revitalizing the habitats as was reported by Glińska-Lewczuk (2004b). When the river water directly supplies an oxbow lake, especially a closed one, it simultaneously regulates its oxygen conditions. It does not allow for oxygen deficiency or an excess of the gas.

The oxbow lake (S4), supplied by the ditch RS4, showed higher indices of correlation with the River Łyna in the case of EC (r = 0.73) and TDS (r = 0.48) than with the ditch water (r = 0.50 and r = 0.44, respectively). A stronger relationship was noted for TP between oxbow lake and ditch water (r = 0.54) than with the river (r = 0.45). However, no statistically significant correlations were found between S4 and the River Łyna for DO, $P\text{-}PO_4$, $N\text{-}NO_3$ and $N\text{-}NO_2$, but it was found between S4 and the ditch water for these compounds (r = 0.48, r = 0.51, r = 0.44, and r = 0.47, respectively). The effect of ditch water RS2 on the oxbow lake S2 is less significant, although concentrations of substances are similar to RS4 water. It can be explained by only a three times higher outflow volume in comparison to the capacity of S2.

Accumulation rates in oxbow lake water

Each of the investigated oxbow lakes retained on average 3.7 tonnes of dissolved solids, with a range from 0.54 for S7 to 8.59 for S3, when the volume of water was referred to mean water level (100 cm on the gauge). The average accumulation of TDS in a water column was 513 g m^{-2}, 71% of which was MM (Table 4). The main threat to the oxbow lake existence is the overload with TP (on average 1.93 g m^{-2} but 3.42 g m^{-2} in the "closed" group) and TIN (on average 1.71 g m^{-2} but 2.85 g m^{-2} in the "open" group) which considerably exceeded Vollenveider's (1968) criteria for dangerous loads for lakes and rivers.

Statistically significant differences in loads among three groups of oxbow lakes were found for all chemical elements studied except for $N\text{-}NH_4$. The maximal loads of most components were found in the open type of oxbow lakes. The load of $N\text{-}NO_3$ is more than seven times higher than in the closed type and two times higher than in the semi-open type. Closed oxbow lakes, in comparison to the other two types, accumulated the highest amounts of TP and $P\text{-}PO_4$, respectively: 3.42 and 2.26 g per 1 m^2 of water table area. This means that TP is retained twice as much as in the semi-open type, and 1.5 times more than in the open type. Differences in $P\text{-}PO_4$ are more distinct: open and semi open types of oxbow lakes retained by four times lower amounts of phosphates than the closed type.

The contribution of the imported material exceeded by almost 16 times that retained by the oxbow lake water when expressed with TDS (Table 5). The highest ratio of ditch input (RS4) to accumulation in oxbow lake (S4) water was found for TIN which was 30 times higher and TP was 22 times higher.

The difference in the reaction of oxbow lake water (S4) to the input of TP and $N\text{-}NO_3$ loads with ditch water is considerable (Fig. 3). TP supplied to the oxbow lake caused a rapid increase in the phosphorus load in oxbow lake water regardless of the season of the year. In the case of $N\text{-}NO_3$, the relationship is more complicated but statistically significant ($r = 0.61$, at $P \leq 0.05$). Intensive export of $N\text{-}NO_3$ from the ditch basin in the period of limited uptake by plants (autumn, mid-winter thawing as well as ice melting in early spring) results in a characteristic decrease in nitrate

Table 4 Accumulation of chemical elements in g per 1 m^2 of oxbow lake water table area. Various letters indicate groups of means different in Duncan test at $P \leq 0.05$.

Type of oxbow lake	Para–meter	TDS	MM	OM	$P\text{–}PO_4$	TP	$N\text{–}NO_3$	$N\text{–}NH_4$
All objects	Mean	513	353	142	0.73	1.93	0.80	0.91
Open	Mean	787^b	522^b	224^b	0.71^a	2.41^{ab}	1.39^b	1.46
	Min–Max	494–1564	220–1162	96–376	0.09–3.61	0.15–1.0	0.02–6.87	0.03–10.57
Semi–open	Mean	405^a	291^a	111^a	0.61^a	1.59^a	0.60^{ab}	0.71
	Min–Max	210–1256	130–959	22–363	0.08–3.22	0.17–12.00	0.01–5.76	0.01–9.12
Closed	Mean	420^a	239^a	122^a	2.26^b	3.42^b	0.19^a	0.52
	Min–Max	234–872	167–414	68–184	0.44–4.50	1.06–6.50	0.10–0.50	0.28–0.85

Table 5 Comparison of chemical elements retained by S4 oxbow lake water and the annual load supplied in RS4 ditch water.

Compound	Average accumulation in S4 oxbow lake water L_{oxl}		Annual load by ditch RS4 L_d		L_d/L_{oxl}
	kg reservoir^{-1}	g m^{-2}	kg reservoir^{-1}	g m^{-2}	
TDS	3565	289	55010	4584	15.9
MM	2504	203	35917	2993	14.7
OM	1069	87	18204	1517	17.4
TP	11.1	0.90	236	19.7	21.9
TIN	9.8	0.79	276	23.0	29.1

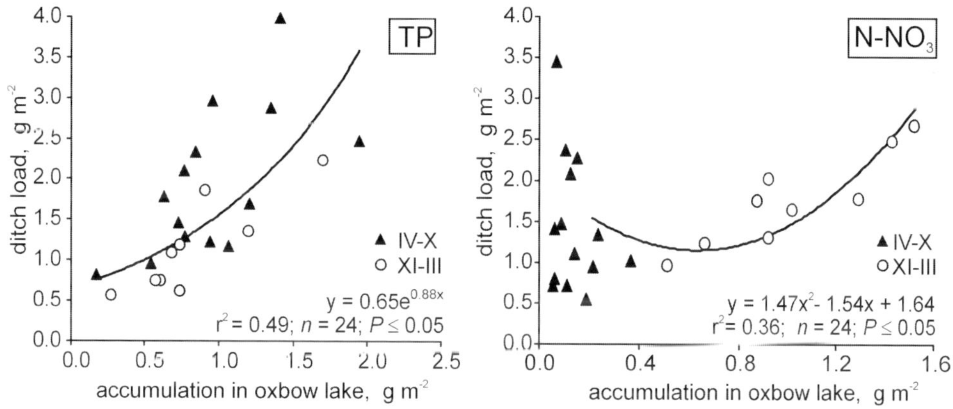

Fig. 3 Relationship between loads of TP and N-NO_3 supplied with RS4 ditch water to S4 oxbow lake and its accumulation in the oxbow lake water.

accumulation in oxbow lake water. After the beginning of the vegetation season (May) there is a proportional increase in the amount of supplied N-NO_3 and accumulation of this compound by the oxbow lake water.

DISCUSSION AND CONCLUSIONS

Oxbow lakes, owing to their position between terrestrial ecosystems and the river, regulate the transfer of matter in the flood plain. This functional attribute of oxbow lakes within the middle sector of the River Łyna valley in northeastern Poland is directly linked to hydrological connectivity as well as to direct input of water from reclaimed areas. Based on the theory of Amoros & Bornette (2002) that connectivity controls nutrient cycling and alters the production and transport phases, an answer was sought to the question of heterogeneity in the water chemistry of the study area. The data obtained showed that oxbow lakes water differs not only from other aquatic habitats over the flood plain area (river water, reclamation ditch) but also among the three groups of oxbow lakes (open, semi-open and closed) selected with regards to retention time and method of matter cycling. The results showed that the nutrient content in oxbow lakes generally increases with the connectivity since the river inputs nutrient-rich water. There are two exceptions to the rule: TP and P-PO_4 in which concentrations and accumulations in closed oxbow lakes were many times higher than in the other two groups. Phosphorus easily undergoes resuspension, and in polymictic shallow water of oxbow lakes causes the secondary pollution with P of the whole mass of water. Closed oxbow lakes are particularly susceptible to degradation of the whole system due to severe oxygen deficiency and increased sediment growth.

Regardless of the type, functionality of oxbow lakes depends on the hydrological character of the river, which in turn regulates the quantity and quality of their water resources. During high water periods, the input of nutrients and dissolved matter dominated the oxbow lake systems. When the flow through the open oxbow lake is maintained and the water is permanently exchanged, more stable oxygen conditions are provided. Increasing water exchange rate (water age) led to a rapid decline of nutrients ($P\text{-}PO_4$) and particulate matter, which was also reported by Hein *et al.* (2004). During long stagnant periods at low water table levels, nutrients in oxbow lakes showed lower concentrations as has been reported by other researchers (Tockner *et al.*, 1999) who attributed it to sedimentation of autochthonous material and nutrient uptake by aquatic organisms.

The original natural character of the investigated oxbow lakes may be easily altered to hypertrophic status due to the agriculture practices within the flood plain and upland areas. When an uncontrolled supply of nutrients (external loading) is added to the existing internal release, the oxbow lake ecosystem may undergo a rapid increase in primary production. As was reported in this study, annual ditch supply of TDS may exceed amounts accumulated in oxbow lake water by on average 16 times; TIN by almost 30 times and P by 22 times.

The natural function of oxbow lakes located in the pathways between reclamation ditches and the river is the regulation of nutrient transfer towards rivers, mainly by retention of matter and, in consequence, prevention of river water quality decline. The interest in the functioning of oxbow lakes as a sink for nutrients in river corridors may be considered, similarly to wetlands, from a possible water purification function (Kadlec & Knight, 1996). However, the direct input of nutrient-rich water to oxbow lakes undoubtedly leads to the deterioration of their trophic status. In order to stop the development of water blooms and to restore the lakes to reasonably healthy conditions, both the external and internal loading has to be reduced. The solution should be found in a river, which is a primary, but to date unexploited, instrument in renaturization of degradated flood plain reservoirs.

Acknowledgements I would like to acknowledge the assistance of S. Kobus and M. Sidoruk with fieldwork, and other participants of the project for their laboratory assistance. I am particularly grateful to Prof. Jozef Koc for inspiration and valuable comments. The research is supported by Polish Scientific Committee grant no. 3PO6S00424.

REFERENCES

Amoros, C. & Bornette, G. (2002) Connectivity and biocomplexity in waterbodies of river flood plains. *Freshwater Biol.* **47**, 517–539.

Amoros, C., Bornette, G. & Henry, C. P. (2000) Environmental auditing—a vegetation-based method for ecological diagnosis of river wetlands. *Environ. Manage.* **25**(2), 211–227.

Glińska-Lewczuk, K., Kobus, S. & Sidoruk, M. (2004a) *Charakterystyka morfometryczna starorzeczy w dolinie środkowej Łyny (Morphometry of Oxbow Lakes in the Middle River Łyna Valley).* Przegląd Naukowy SGGW Warszawa, Poland (in Polish) (in press).

Glińska-Lewczuk, K. (2004b) *Wpływ warunków hydrologicznych na funkcjonowanie starorzeczy w dolinie Łyny (Impact of Hydrological Conditions on Oxbow Lake Functioning in the River Łyna Valley).* Roczn. AR w Poznaniu, Poland (in Polish) (in press).

Glińska-Lewczuk, K. (2004c) Walory przyrodnicze starorzeczy w dolinie środkowej Łyny. (Natural values of oxbow lakes in the middle part of the River Łyna valley). *Mat. Konf. Bliskie naturze kształtowanie dolin rzecznych* 7–9 VI. Rajgród, 1–8 (in Polish).

Hein, T., Baranyi, C., Reckendorfer, W. & Schiemer, F. (2004) The impact of surface water exchange on the nutrient and particle dynamics in side-arms along the River Danube, Austria. *Sci. Tot. Environ.* **328**, 207–218.

Hermanowicz, W., Dożańska, W., Dojlido, J. & Koziorowski, B. (1999) *Fizykochemiczne badania wody i ścieków.* (Physicochemical Methods of Water and Wastewater Analyses). Arkady, Warszawa, Poland (in Polish).

Kadlec, R. H. & Knight, R. L. (1996) *Treatment Wetlands.* Lewis Publishers, CRC, New York, USA.

Tockner, K., Pennetzdorfer, D., Reiner, N., Schiemer, F. & Ward, J. V. (1999) Hydrological connectivity, and the exchange of organic matter and nutrients in a dynamic river-flood plain system (Danube, Austia) *Freshwater Biol.* **41**, 521–535.

Vollenveider, R. A. (1968) Scientific fundamentals of lake and stream eutrophication with particular reference to phosphorus and nitrogen as eutrophication factors. Technical Report DAS/DSI/68.27. OECD Paris, France.

Ward, J. V., Tockner, K., Arscott, D. B. & Claret, C. (2002) River landscape diversity. *Freshwater Biol.* **47**, 517–539.

Webb, B. W., Phillips, J. M. & Walling, D. E. (2000) A new approach to deriving "best-estimate" chemical fluxes for rivers draining the LOIS study area. *Sci. Total Environ.* **251/252**, 45–54.

3 River Corridors, Riparian Zones and Wetlands

Seasonal links between the Amazon corridor and its flood plain: the case of the várzea of Curuaí

M. P. BONNET[1], G. BARROUX[1], P. SEYLER[1], G. PECLY[2], P. MOREIRA-TURCQ[3], C. LAGANE[1], G. COCHONNEAU[4], J. VIERS[1], F. SEYLER[1] & J. L. GUYOT[5]

1 *IRD UR 154, Laboratoire des Mécanismes de Transfert en Géologie, F-31400 Toulouse, France*
bonnet@lmtg.obs-mip.fr

2 *COPPE/UFRJ, Laboratório de Traçadores, 21945-970 Rio de Janeiro, Brazil*

3 *IRD-LMTG -Departamento de Geoquímica, Universidade Federal Fluminense, Morro do Valonguinho s/n, 24020-007 Niterói, RJ, Brazil*

4 *IRD LMTG Maison de la télédétection, F-3400 Montpellier, France*

5 *IRD-LMTG, Casilla 18-1209, Lima 18, Peru*

Abstract Due the large extent (about 600 000 km^2) of the Amazonian flood plain it is expected to greatly influence the Amazon River dynamics, in terms of suspended solids and chemical fluxes. However, up to now, fluxes and exchanges between the main stream and its flood plain are poorly known and it is still difficult to precisely quantify to what extent the flood plain may play a significant role. For several years, in the framework of the HyBAm (Hydrology and Geochemistry of the Amazonian Basin, IRD-CNPq) research programme, *in situ* hydrological and geochemical conditions have been investigated in a flood plain located in the lowest part of the river course. This data set enabled development of a model describing fluxes exchanged between the flooded area and the Amazon River. The model—in a relatively simple form—was in good agreement with the data. The results show that the flood plain is a source of water and of redox sensitive elements, such as arsenic, for the Amazon River.

Key words Amazon River; flood plain; modelling

INTRODUCTION

The Amazon basin drains a surface area of around 6 000 000 km^2, constituting ~5% of the Earth's terrestrial surface. In the middle and lower part of their courses, the Amazon River and its large tributaries are accompanied by large flooded areas named "várzeas" in Brazil. These systems cover a total area of about 300 000 km^2 (Junk, 1997), including 110 000 km^2 along the main stream (Melack & Fisher, 1990) and are composed of permanent and temporary lakes which increase in size and become connected to each other during periods of high water level stage. Due to their large extent, these areas are expected to play a significant role in the river dynamics (Junk *et al.*, 1989; Sparks *et al.*, 1990), such as enabling damping of the hydrograph with efficiency, depending on their storage capacity. Based on a Muskingum formula, Richey *et al.* (1989) indirectly estimated that 30% of the Amazon River flow is routed through the flood plain along a 2000 km reach between Sao Paulo de Olivença and Óbidos. However, the role of the flood plain in fluxes of suspended solids and the biogeochemistry dynamics of the Amazon River is still not yet well quantified.

The major part of the suspended solids in the Amazon River originate from the Andean chain; the particles are strongly altered by deposition/re-suspension cycles in channels and flood plain (Quay *et al.*, 1992). More than 80% of the suspended solids entering the várzeas are deposited (Mertes *et al.*, 1996; Dunne *et al.*, 1998), but the storage in the flood plain is temporary. Dunne *et al.* (1998) estimate that on an annual timescale roughly 2 billion tons of suspended solids enter the flood plain, while 1.5 billion tons are exported from these systems due to re-suspension processes, wind action or erosion. Similarly, flood plains are expected to play a significant role in the carbon cycle. A high primary production takes place in the flood plain (estimated to 115 t C ha^{-1} $year^{-1}$; Junk, 1985) of which 73% is due to phytoplankton and macrophytes. Part of the organic carbon produced in the flood plain is exported and sustains heterotrophic activity in the Amazon River (Quay *et al.*, 1992). Redox conditions present in the flood plain sediment and eventually in the water column may induce dissolution of oxide and hydroxide minerals. A recent study of geochemistry of the Amazon River has shown that some trace elements, such as Mn or Fe, show a concentration maxima a few weeks after the discharge maxima. This was interpreted as the influence of dissolved substances fluxes exported from the várzea in which biochemical processes were favourable to dissolution or desorption (Seyler & Boaventura, 2001, 2003).

In this study, we aim to quantify the exchanged water fluxes as well as the fluxes of some dissolved substances between the Amazon River and a particular várzea. Our approach is based on a relatively simple model which takes advantage of an extensive database.

STUDY SITE AND AVAILABLE DATA

The study site is the várzea of Lago Grande de Curuaí (between 56.10°W and 55.00°W from upstream to downstream, and 2.3°S and 1.9°S) located in the lowest part of the Amazon River near Óbidos (Para state, Brazil), the last gauging station before the estuary (Fig. 1). This várzea has a flood area varying between 1340 and 2000 km^2 (Kosuth, 2002) according to water level stage. Its watershed covers 3660 km^2 including open waters.

This várzea is formed by several white-water lakes (waters characterized by high suspended sediment loads) and black-water lakes (waters characterized by high concentrations of dissolved humic acids and low concentrations of suspended sediment) interconnected with each other and permanently connected to the Amazon mainstream by small channels.

Since 1999, daily water level was recorded at several locations in the várzea. Discharge, using an ADCP (Acoustic Doppler Current Profiler), was estimated in the main connecting channels between the Amazon River and the várzea. Samples for determination of suspended solids and major and trace elements concentrations were taken three times per month and monthly, respectively. Precipitation was recorded daily at Curuaí.

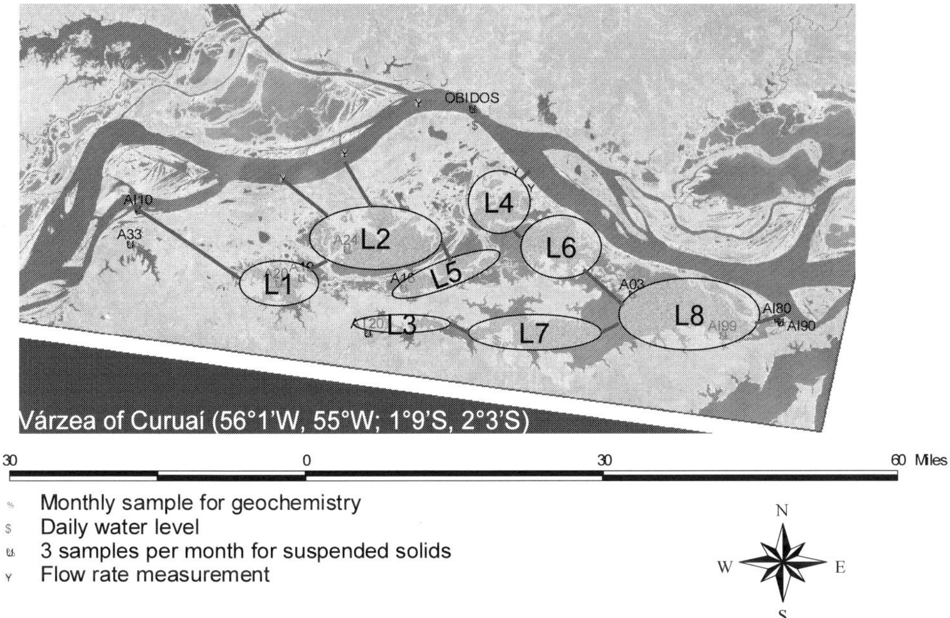

Fig. 1 Map of the site and topographic descriptions used in the model enabling an optimal use of the *in situ* data.

MODEL PRINCIPLE

The objectives of the modelling approach are to quantify the flux of water and some dissolved substances fluxes exchanged between the Amazon River and its flood plain and to determine water renewal time within the different parts of the várzea. The hydrological model is based on three governing equations allowing the description of the water mass balance of the different lakes constituting the várzea. Resolution is achieved using an implicit scheme in time and an upstream scheme in space.

At each time step t:

flow rate Q_{ij}^{t+1} (m^3 s^{-1}) in channel ij linking lake i and j (or Amazon River connecting point i and lake j) is computed by a Strickler-Manning formulation assuming depth $H << L$, where L is the channel width (m).

$$Q_{ij}^{t+1} = K_s \sqrt{p_{ij}^{t+1}} L H_{ij}^{t+1^{5/3}} \quad (1)$$

with K_S the roughness coefficient, $p_{ij}^{t+1} = \dfrac{\left|Z_i^{t+1} - Z_j^{t+1}\right|}{l}$ is the bottom slope (m m^{-1}) assumed equal to the water slope between upstream points Z_i^{t+1} and Z_j^{t+1}. l (m) is the channel length. Flow direction is deduced from the sign of p. $H_{ij}^{t+1} = \dfrac{Z_i^{t+1} + Z_j^{t+1}}{2}$ is the mean depth of the channel.

The water mass balance of each lake B_i ($m^3\ s^{-1}$) is computed as follows:

$$B_i^{t+1} = \left(\sum_{ij} Q_{ij}^{t+1} \right) + Q_{P_i}^{t+1} + Q_{BV_i}^{t+1} \tag{2}$$

where $\sum_{ij} Q_{ij}^{t+1}$ is the algebraic sum of flow rates in ingoing (positive) or outgoing (negative) channels for lake i ($m^3\ s^{-1}$), $Q_{P_i}^{t+1}$ ($m^3\ s^{-1}$) is the flow rate linked to precipitation on S_i^{t+1}, the open water area, whereas $Q_{BV_i}^{t+1}$ ($m^3\ s^{-1}$) is linked to precipitation on the water shed of lake i. Both terms $Q_{P_i}^{t+1}$ and $Q_{BV_i}^{t+1}$ are deduced from daily raingauge data located in Curuaí after subtraction of evaporation. The latter is estimated according to (Riou, 1975) as: $ETP = 0.3T_m - 5.9$, where ETP is the potential evapotranspiration (mm day^{-1}), T_m is the monthly averaged maximum temperature (°C). At this stage, due to a lack of data, the model does not take into account any exchanges with the water table. Additional field investigations will be undertaken in order to investigate these exchanges.

Adjustment of the water level of lake i is then deduced by a simple mass conservation equation :

$$Z_i^{t+1} = Z_i^t + \frac{B_i^{t+1} * \Delta t}{S_i^{t+1}} \tag{3}$$

where S_i^{t+1} is the lake area varying with water level according to a third-order polynomial function deduced from altimetry data (JERS) (Martinez *et al.*, 2003).

Once convergence of the hydrologic model is achieved, the model also allows computation of non-reactive tracer fluxes, assuming a convective transport.

At each time-step t + 1, the fluxes balance in each lake i $B\Phi_{ij}^{t+1}$ ($g\ m^{-3}\ s^{-1}$) is computed according to (4):

$$B\Phi_i^{t+1} = \left(\sum_{ij} \Phi_{ij}^{t+1} \right) + \Phi_{P_i}^{t+1} + \Phi_{BV_i}^{t+1} \tag{4}$$

where $\Phi_{P_i}^{t+1} = Q_{P_i}^{t+1} C_p$ is the flux linked to the rainfall, C_p is the tracer concentration in rain ($g\ m^{-3}$). Similarly, $\Phi_{BV_i}^{t+1}$ is the flux associated with the watershed runoff. $\sum_{ij} \Phi_{ij}^{t+1}$ is the algebraic sum of the input and output fluxes computed according to an upstream scheme.

RESULTS AND DISCUSSION

The model has been applied for a three-year period from January 2001 to December 2003. At this stage, eight interconnected lakes are taken into account as shown in Fig. 1. This topographic description allowed advantage to be taken of the *in situ* collected data

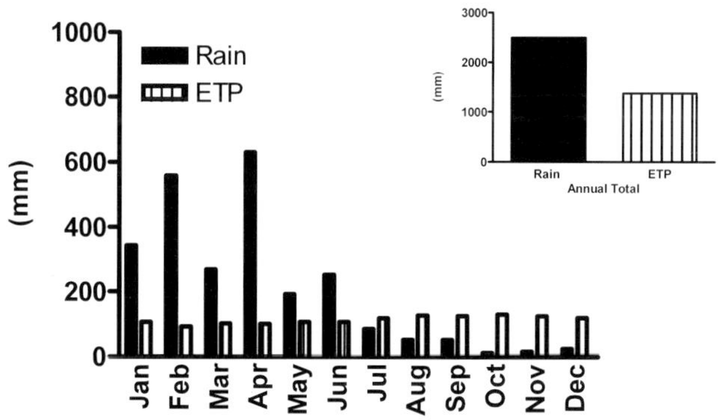

Fig. 2 Measured precipitation and estimated evapotranspiration in Curuai during 2001.

that show that the flood plain is not a homogeneous system, and was necessary in order to give an estimation of the water renewal time in the different lakes of the várzea.

Data required for simulation are evaporation and precipitation. As shown in Fig. 2, during year 2001, evaporation accounts for roughly 55% of the rainfall. Highest precipitation occurs from January to April, while evapotranspiration is nearly constant throughout the year. Besides daily precipitation and evapotranspiration, data requirements for the simulation include water level measured in the Amazon River at the mouths of different connecting channels. This data is provided by the national water agency (ANA, Brazil) at three locations, Parintins (40 km upstream from the várzea), Óbidos and Santarem (40 km downstream from the várzea). Linear interpolation is used to determine the water level at the relevant locations. In addition, an initial water elevation is required for all the lakes.

The model was calibrated against water level measured in Curuaí and discharges measured at the different várzea–Amazon River connecting channel mouths. The parameters that require calibration are related to channel geometry (width, depth, length) and the roughness coefficient, *Ks*. Among all these parameters, the latter is the least sensitive; it has been fixed to 40 for all channels. Channel length was deduced from satellite imagery and field data, whereas channel depth and width were mostly deduced from current profiles acquired during the field campaigns.

The simulated water level shows a very good agreement with measured water level (Fig. 3), the mean absolute difference does not exceed 7 cm and is similar for each simulated year of the study (calibration period (2002) and the validation period (2001 and 2003)). Moreover, for most of the várzea–Amazon River channels, the computed flow rate is in good agreement with the data acquired during the field campaigns (Fig. 4). However, additional discharge measurements would be necessary to establish realistic rating curves of the different connecting channels in order to get more confidence in the model validation. As seen from Fig. 4, upstream channels AI 10 to AI 30 only operate as a source of water for the várzea, whereas the downstream channel AI80 mainly behaves as a sink.

During an annual hydrological cycle, the várzea behaves as a source of water for the Amazon River, but exported fluxes only account for roughly 0.45% of the annual average water flux of the Amazon River estimated at Óbidos (the mean average flow

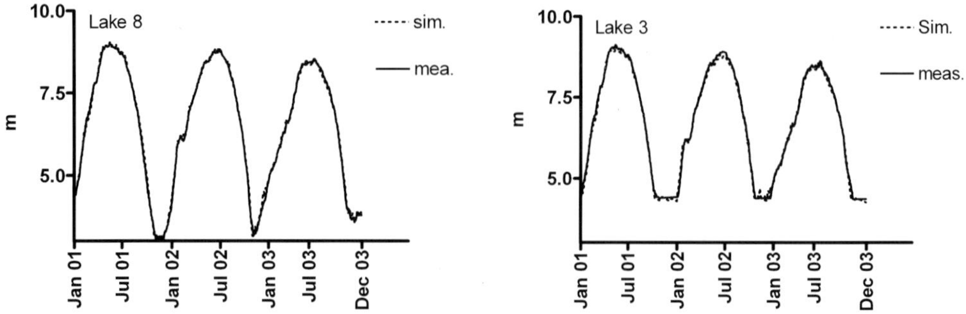

Fig. 3 Comparison of the measured and simulated water levels in the eastern part of the várzea (Lake 8) and the western part (Lake 3) of the várzea.

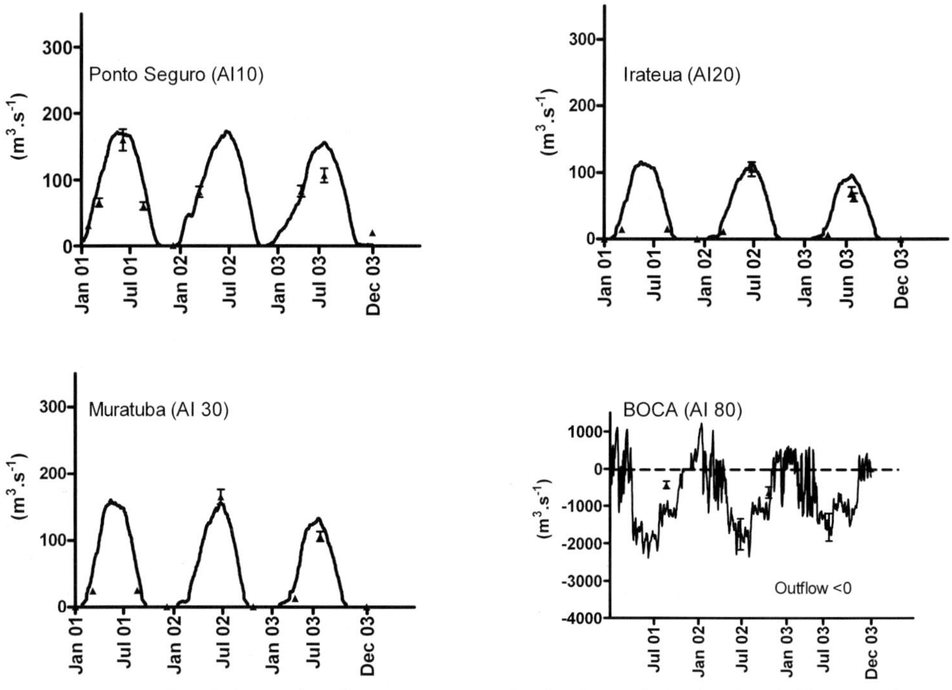

Fig. 4 Comparison between measured (triangle) and simulated (solid line) discharges in different connecting channels.

being 169 000 $m^3 s^{-1}$). However, assuming a similar behaviour for all the flood plains in the Basin (300 000 km^2), we find that about 35% of the Amazon River and major tributaries is routed through these systems. This value is very close to that found by Richey *et al.* (1989) as mentioned in the introduction.

Once the exchanges with the Amazon River were calibrated, the model was applied to describe the transport of nonreactive tracers. This step was necessary to calibrate the exchanges between the different parts of the várzea, in order to give insights on the renewal time in the different lakes constituting the study area. The results of chlorine simulation for the two main lakes of the várzea are shown in Fig. 5.

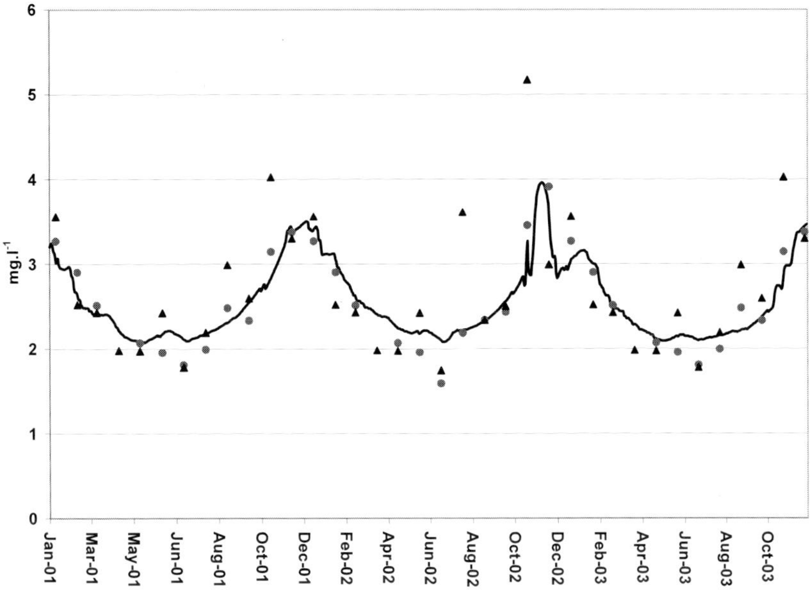

Fig. 5 Comparison of the measured (circle) and simulated (solid line) chlorine concentration in the eastern part of the várzea. Chlorine concentration in the Amazon River is also reported (triangle).

A good agreement between the simulation and the data is obtained for lake 8, which corresponds to Lago Grange of Curuaí in the eastern part of the várzea. This lake is very well connected with the Amazon River as shown in Fig. 1. Temporal variation of its chlorine concentration is mainly driven by that of the Amazon River, with a small dilution by rain. Once more, some additional calibration work is required in order to simulate the temporal variations in the other parts of the várzea properly. Moreover, the model shows large heterogeneities in the water residence time from lake to lake, with a minimum of less than 1 month and a maximum of more than 6 months. The mean water renewal time is about 3.5 months which is in good agreement with value proposed by Kosuth (2002).

As mentioned by Seyler & Boaventura (2003), redox conditions prevailing in the várzea may induce dissolution or desorption of some trace elements. In particular, it is expected that the várzea behaves as a net source of elements such as arsenic (As), iron or manganese. In order to verify this assumption, we used the model to compute exchanged fluxes of As, using the measured concentration in the eastern part of the várzea. Results of the model confirm that the várzea behaves as a source of As (Fig. 6). Export occurs mainly during the decreasing water level stage (May–October) and is maximal during the low water level period. The exported flux from the várzea to the Amazon River accounts for only about 0.3% of the annual average As flux at Óbidos. Using a similar approach as used for water fluxes (see above) to extrapolate to all the flood plains, we find that these systems could account for 24% of the As flux at Óbidos. These results are in good agreement with Seyler & Boaventura (2003). The highest export of As, especially during low water periods, can be explained by the increase of the water residence time and by reductive conditions at the sediment–water interface.

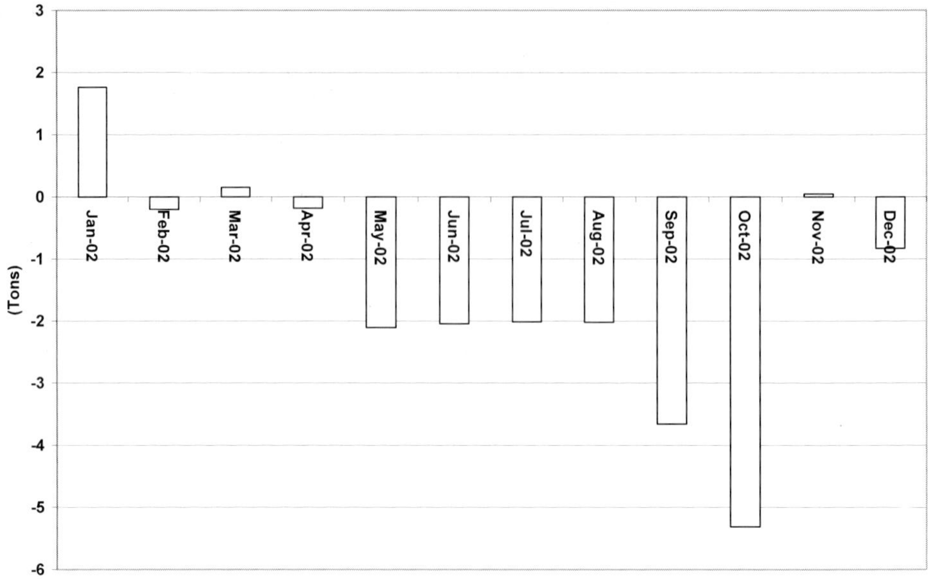

Fig. 6 Exchanged arsenic mass between the várzea and the Amazon River. Negative values indicate exportation from the várzea.

CONCLUSIONS

In this study, we propose a model to quantify exchanged fluxes between the Amazon River and one of its flood plain lakes systems. Despite a relatively low level of complexity, the model showed a good agreement with the *in situ* data. The várzea behaves as a source of water for the Amazon River. Extrapolating to similar hydrosystems, we found that the whole flood plain in the Amazonian Basin may account for 35% of the mean annual flux at Óbidos. Using the model to quantify the exchanged flux of arsenic, a redox sensitive element, we showed that the várzea behaves as a source of such types of element. Extrapolation to the whole Amazon River flood plain indicates that such systems could account for 24% of the mean annual As flux at Óbidos. The computed residence time for the whole várzea is about 3.5 months but large heterogeneities are expected from lake to lake, from less than 1 month to more than 6 months. However, the latter result needs to be confirmed by additional calibration work.

REFERENCES

Dunne, T., Mertes, L. A. K., Meade, R. H., Richey, J. E. & Forsberg, B. R. (1998) Exchanges of sediment between the flood plain and channel of the Amazon River in Brazil. *Geol. Soc. Am. Bull.* **110**, 450–467

Junk, W. J. (1985) The Amazon floodplain: a sink or source for organic carbon? In: *Transport of Carbon and Minerals in Major World Rivers*, 267–283 Pt3. SCOPE/UNEP.

Junk, W. J. (1997) *The Central Amazon Flood Plain: Ecology of a Pulsing System*. Ecological Studies 126. Springer-Verlag, Berlin, Germany.

Junk, W. J., Bayley, P. B. & Sparks, R. E. (1989) The flood pulse concept in river-floodplain systems. (In: Proc. Int. Large River symp.) *Can. Special Publ. Fish Aquat. Sci.* **106**, 110–127

Kosuth, P. (2002) Water and sediment dynamics in the Várzea of Lago Grande de Curuaí. In: *Second International LBA Scientific Conference*) Manaus, 7–10 July 2002.

Martinez, J. M., Kosuth, P., Cochonneau, G., Guyot, J.-L., Seyler, F., Bourrel, L. & Maurice-Bourgoin, L. (2003) Application of remote sensing data for the quantification of an Amazon floodplain extension, dynamic and river water storage. In: *Proc. Joint Conference of European Geophysical Union and American Geophysical Union* (Nice, France, Avril 2003).

Quay, P. D., Wilbur, D. O., Richey, J. E., Hedges, J. I., Devol, A. H. & Victoria, R. (1992) Carbon cycling in the Amazon River: implication from the ^{13}C compositions of particles and solutes. *Limnol. Ocean.* **37**(4), 857–871.

Richey, J. E., Mertes, L. A. K., Dunne, T., Victoria, R. L., Forsberg, B. L., Tancredi, A. C. N. S. & Oliveira, E. (1989) Sources and routing of the Amazon River flood wave. *Global Biogeochem. Cycles* **3**, 191–204.

Riou, C. (1975) La détermination pratique de l'Evaporation. Application à l'Afrique centrale. Thèse. Mémoires Orstom no. 80. Orstom, Paris, France.

Seyler, P. & Boaventura, G. R. (2001) Trace elements in the mainstream Amazon River. In: *The Biochemistry of the Amazon Basin* (ed. by M. E. McClain, R. L. Victoria & J. E. Richey), Chap. 16. Oxford University Press, Oxford, UK.

Seyler, P. & Boaventura, G. R. (2003) Distribution and partition of trace metals in the Amazon Basin. *Hydrol. Processes.* **17**, 1345–1361.

Dynamics and Biogeochemistry of River Corridors and Wetlands (Proceedings of symposium S4 held during the Seventh IAHS Scientific Assembly at Foz do Iguaçu, Brazil, April 2005). IAHS Publ. 294, 2005.

The influence of spatial vegetation distribution on Taim Wetland hydrodynamics

ADRIANO ROLIM DA PAZ, ADOLFO NICOLÁS VILLANUEVA & EDITH BEATRIZ CAMAÑO SCHETTINI

Instituto de Pesquisas Hidráulicas, Universidade Federal do Rio Grande do Sul (IPH/UFRGS), Av. Bento Gonçalves, 9500, CEP 91501-970, Porto Alegre (RS), Brazil
adrianorpaz@yahoo.com.br

Abstract Water circulation plays an important ecological role in ecosystems such as wetlands. Thus, understanding wetland hydrodynamics becomes an essential factor to the success of conservation and restoration projects. This paper describes the application of a two-dimensional, depth-averaged, hydrodynamic model to the system formed by Lake Mangueira and the Taim Wetland (Brazil). Vegetation resistance to flow was represented in a modelling system using a Manning roughness coefficient higher than that in unvegetated areas. The wind protection effect of vegetation was modelled by decreasing water surface shear stress in vegetated areas. Three spatial vegetation distributions were considered, resulting in very different water circulation patterns in the Taim Wetland. The influence of Lake Mangueira on the Taim Wetland was found to be completely dependent on vegetation occurrence in the interface between these ecosystems. Simulation results illustrate how spatial vegetation distribution changes may alter Taim Wetland hydrodynamics, having potentially detrimental ecological effects on the entire ecosystem.

Key words Taim Wetland; two-dimensional hydrodynamic modelling; vegetation resistance

INTRODUCTION

Although it is not yet well recognized, water circulation is a fundamental factor in the establishment of wetland structure and functioning (Aumen, 2003). Hydrodynamics influences ecological patterns such as distribution and diversity, as well as the amount of vegetation, invertebrates and fish that are dependent on aquatic plants, and thus plays an essential role in the development of restoration and re-naturalization projects (Newall & Hughes, 1995). In turn, changes in spatial vegetation distribution may alter water circulation patterns, constituting an interdependent relationship (Wheeler & Shaw, 1995).

Vegetation influences hydrodynamics in two ways: (a) it offers an additional resistance to flow over the entire water column (in the case of emergent plants) or in part of it (submerged plants); (b) it protects the water surface against wind shear stress, and can even exclude wind action as a flow-governing force when there is a dense plant canopy (Kadlec, 1990).

This paper analyses the impact of different hypothetical spatial vegetation distributions on the hydrodynamics of the Taim Wetland (a natural wetland in southern Brazil), using a two-dimensional (2-D), depth-averaged, hydrodynamic model coupled with a wetting and drying algorithm. The main objective of this research was to use numeric simulations to estimate how possible spatial vegetation distribution changes in Taim Wetland influence its hydrodynamics.

This research is associated with the PELD/CNPq program (Long Term Ecological Research Program, Taim Hydrological System), which is included in a national network created to understand the structure and functioning of these ecosystems and determine ecological patterns to produce knowledge necessary for wetland management.

TAIM WETLAND

The Taim Wetland is included in the Taim Hydrological System, which is part of the Taim Ecological Station (ESEC Taim), a Brazilian environmental protection area created by Decree no. 92.963 (21 July 1986), with the purpose of protecting southern wetland sample areas, as well as species on the verge of extinction and resting areas for migratory birds. When created, ESEC Taim had an area of 315 km^2 located in the southern part of the state of Rio Grande do Sul (Fig. 1). In June 2003, a new decree expanded the station area to 1100 km^2, including the entire area between Lake Mangueira and the Atlantic Ocean.

Rice farming constitutes the major economical activity in the region and is virtually the only recourse of employment for the local population. Water withdrawal from Lake Mangueira for this activity has altered hydrological patterns in the system formed by the lake and Taim Wetland (Villanueva *et al.*, 2000). Besides the conflict

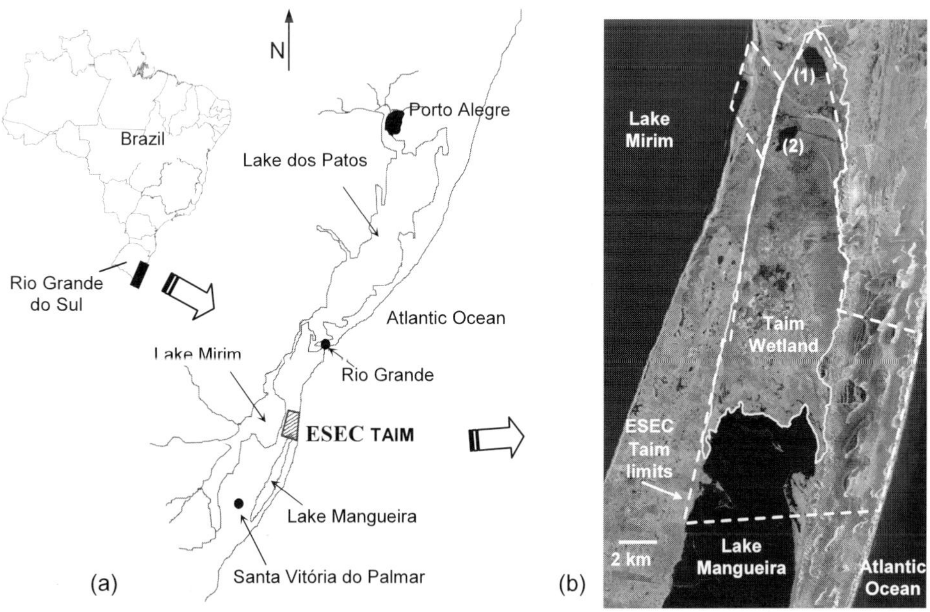

Fig. 1 (a) Taim Ecological Station in southern Brazil; (b) satellite image of ESEC Taim limits and Taim Wetland limits, where (1) and (2) indicates Lakes Nicola and Jacaré.

between rice farmers and ecosystem conservation interests, there are other human impacts, such as chemical and agrotoxic pollution of the lake, predatory fishing and hunting, as well as animals being run over on Federal Roadway BR-471, which defines the western limit of the wetland.

VEGETATION MAPS

Emergent aquatic macrophytes are the principal vegetation in the Taim Wetland and are dependent on characteristics that enable them to withstand the anaerobic conditions of the soil when there is a certain water level or saturated soil. However, the occurrence of extreme water depths or the absence of a minimal water level over a prolonged period of time can compromise the development of such plants (Marble, 1992).

Assuming that water level fluctuation in Taim Wetland shifts its spatial vegetation distribution patterns, it can be supposed that the water level oscillation around a fixed value over a long enough period of time would result in a particular spatial vegetation distribution. Based on this hypothesis, we considered the occurrence of such an oscillation around three fixed water levels (low, 2.4 m; medium, 2.9 m; high, 3.4 m) established from the historical variation of the Taim water level. For each fixed water level, a map was obtained with a distinct spatial configuration of wetland vegetation (Fig. 2): (a) low level (type A); (b) medium level (type B); (c) high level (type C).

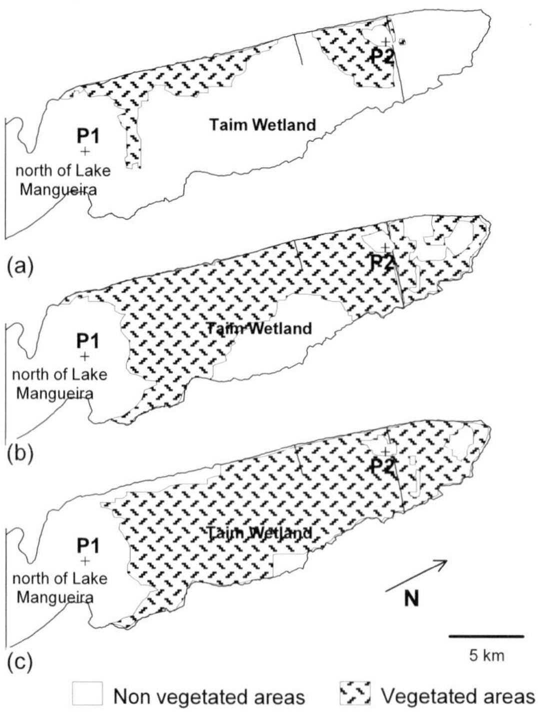

Fig. 2 Type A (a), B (b) and C (c) vegetation maps with indication of points P1 and P2 of the analysis of water level and flow velocity.

Values from 0.1 m to 1.8 m were adopted as the minimum and maximum limits of depth that provide the conditions emergent aquatic macrophytes can withstand. Such values were established based on the literature (Hammer, 1992) as well as observations of the Taim Wetland by Motta Marques (personal communication). It should be pointed out that there are no exact depth limits. There are, however, average values around which field observations have reported a greater frequency or lack of the type of vegetation in question.

SYSTEM MODELLING AND SIMULATION

The Taim Wetland was modelled in conjunction with Lake Mangueira, which is the main influx source for the wetland. It employed the IPH-A computational system developed at the Institute for Hydraulics Research (IPH-UFRGS, Brazil). The IPH-A model is a 2-D, depth-averaged, hydrodynamic model (2DH) that takes into account the Manning (or Chézy) formulation for calculating bottom shear stress, while water surface shear stress is determined as a function of wind velocity and a constant drag coefficient (Casalas, 1996). To enable the model to withstand the emergence of "islands" due to water level fluctuations, i.e. the occurrence of internal dry areas in the modelled region, a wetting and drying algorithm was developed and coupled with a IPH-A model (Paz *et al.*, 2004).

Vegetation resistance was represented by an elevated Manning coefficient (n), following the procedure adopted by Somes *et al.* (1999) and Guardo & Tomasello (1995), as well as by a reduction in the wind shear stress (C_{red}), as performed by Podsetchine & Schernewski (1999) and Ramming (1979). Based on the literature and also on the analysis of the sensitivity response of the modelled system (Paz, 2003), the following values were employed for the above-mentioned coefficients: (a) vegetated areas: $n = 0.4$ and $C_{red} = 0.2$; (b) unvegetated areas: $n = 0.02$ and $C_{red} = 1.0$.

A numeric mesh with a spacing of 360 m was used, carrying out a rotation of coordinates in order to reduce the dimensions of the element matrixes. A 60-s time interval calculation, a turbulent viscosity coefficient equal to 10 $m^2\ s^{-1}$, and a wind adjustment coefficient of 3.0E-6 were also used. These coefficients values were chosen based on the literature and on the analysis of the sensitivity response of the modelled system (Paz, 2003).

Three situations were simulated, varying the spatial configuration of the vegetation alone (Map A, B or C). Based on an analysis of the historical series of winds available for the Taim Wetland region, we adopted a wind with a predominantly southeastern direction, selecting a period of four consecutive days from the historical series. In addition to the wind presenting a southeasterly direction with considerable frequency, this direction is of great interest, as it is aligned with the position of the Lake Mangueira–Taim Wetland system.

RESULTS AND DISCUSSION

As expected, the vegetated regions investigated generally exhibited flow patterns that were quite distinct from the unvegetated regions. As a result, the three maps of hypothetical vegetation led to water circulation patterns that were considerably different from one another.

The modelled oscillation of water level at Lake Mangueira experienced practically no influence from the spatial configuration of the vegetation distribution on the Taim Wetland (Fig. 3). It did, however, appear to play a fundamental role regarding the influence of the lake *over* the wetland. The occurrence of vegetation in the extreme southern region of the Taim Wetland, in the interface between the wetland and Lake Mangueira, considerably reduced the interaction between the two systems and made the velocity fields corresponding to each of the hypothetical vegetation maps more distinct from one another (Fig. 4).

In vegetation map A, only the eastern part of the interface between wetland and lake is not covered by vegetation (Fig. 2), causing the interaction between the wetland and Lake Mangueira to develop principally through this point (Fig. 4(a)). With this spatial vegetation distribution, the existence of a continuous unvegetated area from the interface to the central region of the wetland facilitated the influx of water from the lake. This gave the oscillation of the lake water level influence over a greater extension of the wetland.

In vegetation map B, the entire interface between the wetland and Lake Mangueira is covered by vegetation, constituting a strong resistance to the influx of water from the lake to the wetland (Fig. 4(b)). A slightly more distinct situation was obtained with vegetation map C, which presented a narrow unvegetated strip at the western end of the interface that permitted a restricted influx of water from the lake to the wetland (Fig. 4(c)).

The way the vegetation was spatially distributed in the Taim Wetland had direct repercussions on the hydrodynamic behaviour of the internal lakes (Nicola and Jacaré). When it was considered vegetation covering the entire surrounding area of these lakes, its hydrodynamics was quite independent from other regions of the wetland. Considering vegetation maps B and C, the water level in Lake Jacaré oscillated very little in comparison to the case with vegetation map A, as illustrated in Fig. 5. In the case of maps B and C, the hydrodynamic behaviour of Lake Jacaré was practically limited to water circulation induced by the direct action of the wind on its surface.

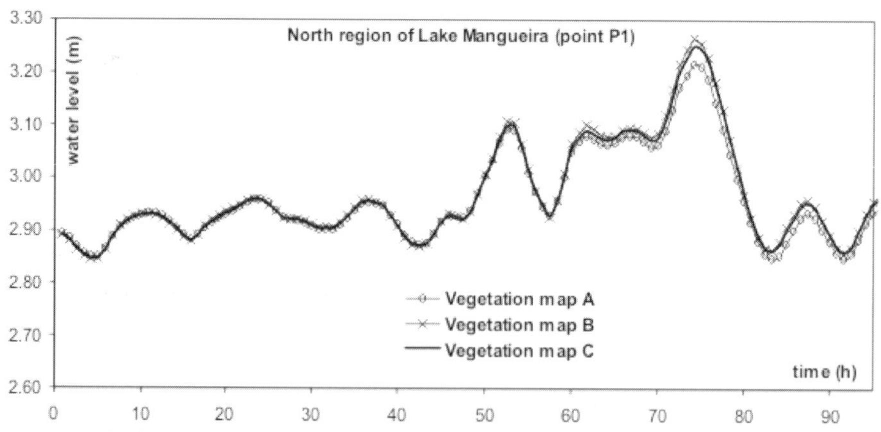

Fig. 3 Oscillation of water level in the extreme northern region of Lake Mangueira for the three simulated cases over time.

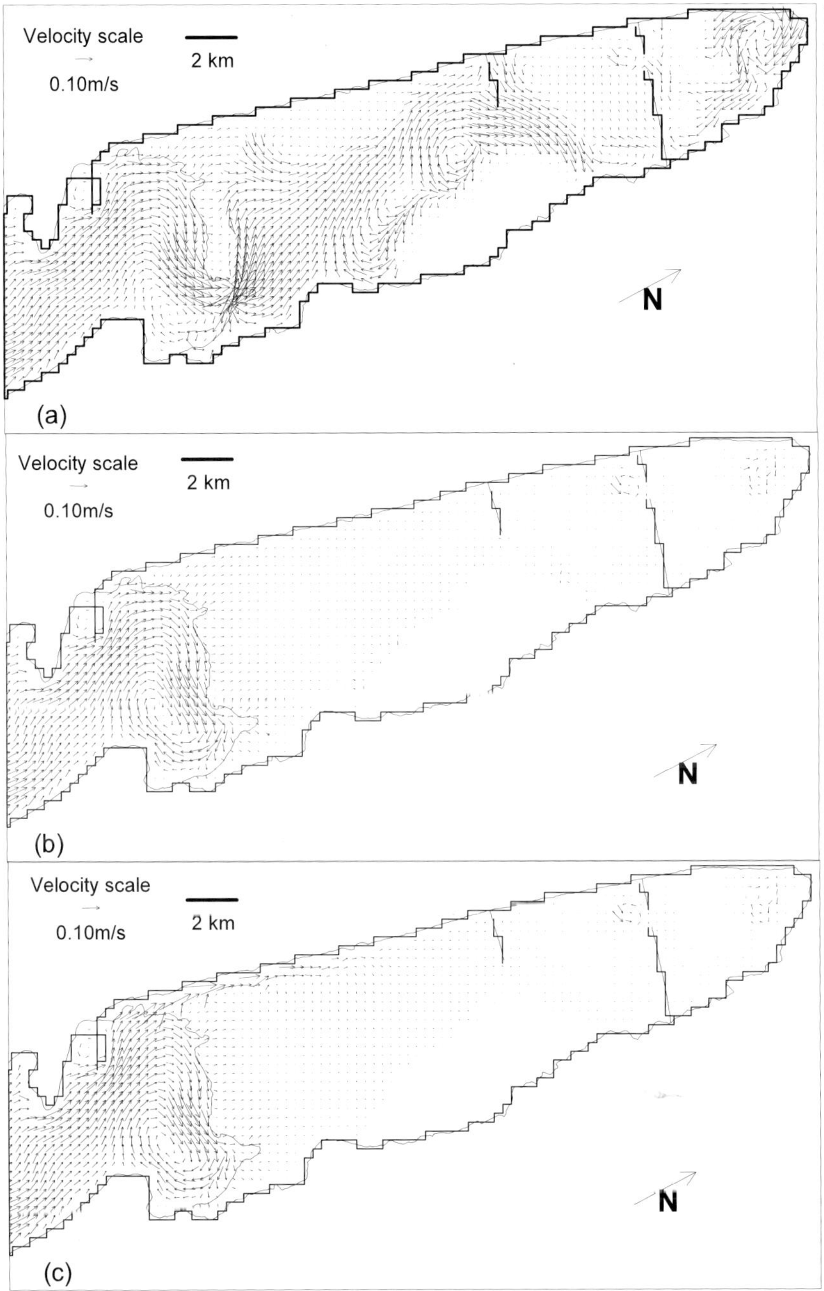

Fig. 4 Velocity fields at time t = 60 h, considering Vegetation Maps A (b), B (c) and C (c).

CONCLUSIONS

In the numeric simulations carried out on Taim Wetland, the vegetated and unvegetated areas exhibited quite distinct hydrodynamic behaviours. Despite the

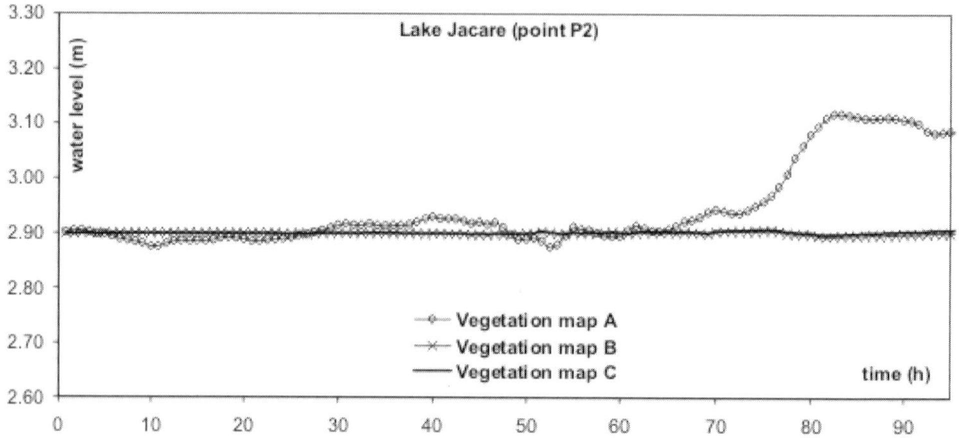

Fig. 5 Oscillation of water level at Lake Jacaré for the three simulated cases over time.

simplification adopted in the present study regarding the representation of the vegetated areas through the employment of an elevated Manning coefficient and a reduced wind shear stress, the distinction between vegetated and unvegetated areas is a factor of great relevance to wetland hydrodynamics and should be considered in the mathematical modelling of such ecosystems.

Lake Mangueira constitutes the main influx of water for the Taim Wetland. Based on the simulated cases, a dependency was verified regarding the influence of the lake over the wetland and the spatial distribution of vegetation in the wetland, especially at the interface between the two systems. The greater presence of vegetation in this region leads to a reduced interaction between the wetland and the lake. Likewise, the existence of a continuous unvegetated area from the interface intensifies the influence of the lake over the wetland and constitutes the main area of influx of water for wetland.

The simulations considered three hypothetical vegetation maps in the Taim Wetland and demonstrated how alterations in the spatial vegetation distribution can impact water circulation patterns as well as the oscillation of water levels, with repercussions on the functioning of the entire ecosystem. This exemplifies the importance of developing research on the hydrodynamics of wetlands, as pointed out by Aumen (2003).

To continue the development of the PELD/CNPq program (Taim Hydrological System), of which the present study is a part, further studies will spatially characterize the occurrence of vegetation in the Taim Wetland. With this, we hope to obtain important information in such a way that modelling comes closer to the actual conditions of the ecosystem. In this way, it will produce valuable information to understand ecological patterns in wetland, e.g. distribution and diversity of fish and invertebrates.

Acknowledgements The authors are grateful to CNPq (Conselho Nacional de Desenvolvimento Científico e Tecnológico, Brazil) for the scholarship of the first author, and to the support from the Long Term Ecological Research Program (PELD/CNPq) to which the present research is linked.

REFERENCES

Aumen, N. G. (2003) *The role of flow in the Everglades—ridge and slough landscape*. South Florida Ecosystem Restoration Working Group, Florida, USA.

Casalas, A. B. (1996) *IPH-A—Aplicativo para modelação de estuários e lagoas, Manual de Uso*. Recursos Hídricos 33, IPH, Porto Alegre, Brazil.

Guardo, M. & Tomasello, R. S. (1995) Hydrodynamic simulations of a constructed wetland in South Florida. *Water Resour. Bull.* **4**(31), 687–701.

Hammer, D. E. (1992) *Creating Freshwater Wetlands*. Lewis Publishers, Boca Raton, USA.

Kadlec, R. H. (1990) Overland flow in wetlands: vegetation resistance. *J. Hydraul. Engng ASCE* **5**(116), 691–706.

Marble, A. D. (1992) *A Guide to Wetland Functional Design*. Lewis Publishers, Boca Raton, USA.

Newall, A. M. & Hughes, J. M. (1995) Microflow environments of aquatic plants in flowing water wetlands. In: *Hydrology and Hydrochemistry of British Wetlands* (ed. by J. M. Hughes & L. Heathwaite), 363–381. John Wiley & Sons, Chichester, UK.

Paz, A. R., Villanueva, A. & Schettini, E. (2004) Wetting and drying algorithm to two-dimensional hydrodynamic modelling of wetlands. In: *XXV CILAMCE* (*Iberian Latin American Congress on Computational Methods in Engineering*). Recife, Brazil.

Paz, A. R. (2003) Análise dos principais fatores intervenientes no comportamento hidrodinâmico do banhado do Taim (RS). MSc Thesis, Instituto de Pesquisas Hidráulicas, Universidade Federal do Rio Grande do Sul, Porto Alegre, Brazil.

Podsetchine, V. & Schernewski, G. (1999) The influence of spatial wind inhomogeneity on flow patterns in a small lake. *Water Res.* **15**(33), 3348–3356.

Ramming, H-G. (1979) The dynamics of shallow lakes subject to wind – an application to Lake Neusiedl, Austria. In: *Hydrodynamics of Lakes* (ed. by W. H. Graf & C. H. Mortimer), 65–75. Developments in Water Sciences 11. Elsevier, Amsterdam, The Netherlands.

Somes, N. L, Bishop, W. A. & Wong, T. H. (1999) Numerical simulation of wetland hydrodynamics. *Environ. Int.* **6/7**(25), 773–779.

Villanueva, A., Motta Marques, D. & Tucci, C. (2000) The Taim Wetland conflict: a compromise between environment conservation and irrigation. *Water Int.* **4**(25), 610–616.

Wheeler, B. D. & Shaw, S. C. (1995) Plants as hydrologists? An assessment of the value of plants as indicators of water conditions in fens. In: *Hydrology and Hydrochemistry of British Wetlands* (ed. by J. M. Hughes & L. Heathwaite), 63–82. John Wiley & Sons, Chichester, UK.

Dynamics and Biogeochemistry of River Corridors and Wetlands (Proceedings of symposium S4 held during the Seventh IAHS Scientific Assembly at Foz do Iguaçu, Brazil, April 2005). IAHS Publ. 294, 2005.

Role of small valleys and wetlands in attenuation of a rural-area groundwater contamination

ADRIAN GALLARDO & NORIO TASE

School of Life and Environmental Sciences, University of Tsukuba, Tsukuba, Ibaraki 305-8572, Japan

adgallardo@yahoo.co.jp

Abstract The transport of nitrate and the role of valley corridors and wetlands in the attenuation of groundwater pollution were evaluated in a small rural catchment in Japan. The site is located at the boundary between uplands and lowlands and is divided into a shallow and a deep aquifer. Shallow waters are rich in nutrients, but concentrations decrease dramatically with depth and towards a wetland. Simultaneous decreases in redox potential and oxygen, and increases in HCO_3^- and pH, suggest denitrification is taking place. Nitrate removal typically peaks within the first few metres of the valleys lowlands and therefore there is a significant potential for NO_3^- reduction within the rest of the lowland buffer strip, and also with depth beneath the uplands. The present study provides a new contribution for understanding the fate of NO_3^- in agricultural areas, and constitutes one of the first works of this type carried out in the region.

Key words agricultural pollution; denitrification; nitrate; wetland

INTRODUCTION

Contamination of shallow aquifers by high levels of NO_3^- has become one of the most common problems in rural regions of Japan, mainly as a result of large-scale fertilizer applications. Several studies have demonstrated that wetlands and riparian strips are able to remove large amounts of nutrients from waters flowing through them (Haycock & Pinay, 1993; Cey *et al.*, 1999). Depletion of NO_3^- with depth also has been reported by some authors (Postma, 1991; Kelly, 1997). However, there is still considerable uncertainty about the exact mechanisms of attenuation, and the processes controlling NO_3^- dynamics cannot be conclusively defined.

In spite of the growing concern about water quality, little work has been done in the Kanto basin, Japan. Thus, the primary objectives of the present study are to determine the spatial and temporal variations of NO_3^- in groundwater in an agricultural catchment of the region, to relate the observations to land use and hydrogeological characteristics, and to assess the role of valley corridors and wetlands in the attenuation process.

FIELD SITE

The study site consists of a catchment of irregular shape, extending approximately 400 × 300 m in the northern suburbs of Tsukuba City, about 60 km northeast of Tokyo, Japan.

The area is located between the uplands of the Tsukuba plateau on the south and the flood plain of the Sakura River on the north. The flat upland presents the maximum elevations, and is connected with the poorly drained and usually swampy lowlands by a steep slope. A few valleys dissect both the slope and lowlands. An artificial canal at the northernmost edge of the area of study drains most of the excess water, while a narrow stream at the western boundary of the site acts as a secondary sink. Land use is dominated by agriculture throughout most of the uplands (Fig. 1). Inorganic fertilizers of [$(NH_4)SO_4$] 14–15% in N, and 45%-N urea are the major sources of NO_3^- to groundwater. Liming with $(CaMg)CO_3$ is also carried out to prevent soil acidification. The rest of the uplands are covered by a dense forest not affected by human activities. The slopes and most of the lowlands are left as marginal lands. A small orchard exists within one of the valleys.

The top of the geological sequence corresponds to the volcanic ash of the Kanto Loam (Table 1), which is underlain by clays of the Jōso Fm. This layer abruptly disappears by erosion at the slope boundary. Below, the sands of the Ryugasaki Fm constitute the upper aquifer of the region. The unit is underlain by a succession of clay

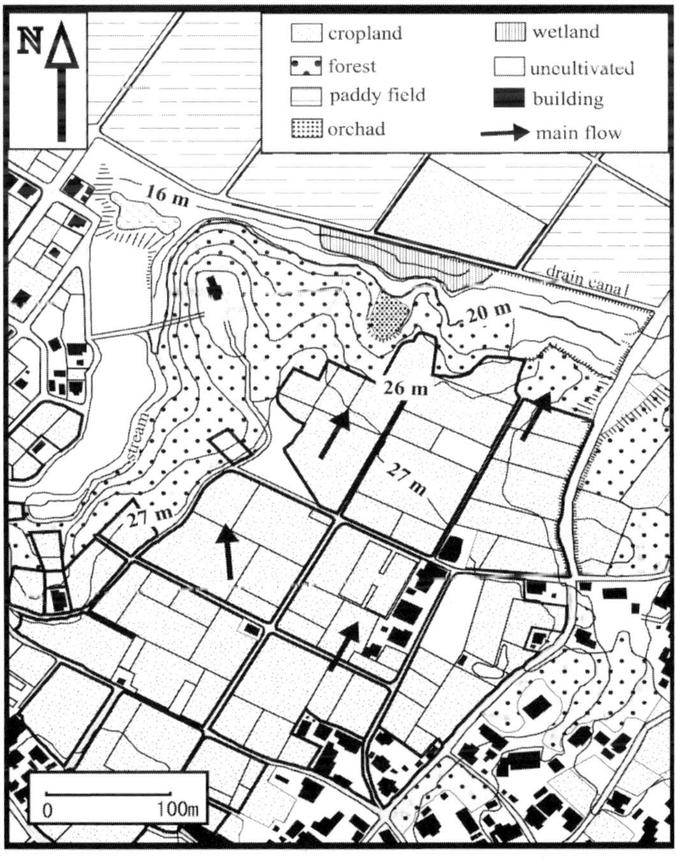

Fig. 1 Land-use in the area of study.

Table 1 Stratigraphy of the site.

Formation	Thickness (m)	Main sediments	Kh ($cm\ s^{-1}$)	Kv ($cm\ s^{-1}$)
Kanto loam	0.75–1.5	Silty clay		2×10^{-4}–8×10^{-5}
Jōso clay	0.7–1.3	Clay		8×10^{-5}–9×10^{-5}
Ryugasaki Fm.	2.8–4.2	Medium-coarse sands & gravels	1×10^{-3}–3×10^{-7}	9×10^{-3}–9×10^{-5}
Narita Fm.	0.5–3.5	Clay. Sand lenses		2×10^{-6}–5×10^{-7}
	1 – >4	Medium-fine sand	2×10^{-5}	2×10^{-5}–1×10^{-4}
	>1.5	Silt	3×10^{-8}	

Kh, horizontal hydraulic conductivity; Kv, Vertical hydraulic conductivity.

and subordinated sand lenses on the uplands, but is in direct contact with a deeper aquifer of the Narita Fm within the lowlands. A nearly impermeable silt layer forms the basement of the sequence. Finally, a horizon highly rich in organic matter locates at the vicinities of the wetland, and in the waterlogged valleys near the drain canal.

METHOD AND MATERIALS

A network of 20 monitoring wells scattered throughout the area, and a transect of 23 multilevel wells along the general groundwater direction constituted the basis to study groundwater and contaminant distribution. Samples were collected monthly for determination of major ions. Dissolved organic carbon (DOC) was analysed for some of the samples, and complemented with measurements carried out by Sugawara (2004). Temperature, conductivity, redox potential (ORP), DO, and pH were measured in the field.

Undisturbed core samples were collected for physical determinations, and horizontal hydraulic conductivity values were estimated by slug test procedures.

The finite-difference program MODFLOW (McDonald & Harbaugh, 1988) was utilized as a complementary tool to quantify groundwater flow and nitrate transport under steady state conditions.

RESULTS

Groundwater flow is essentially from south to north in the uplands, turning to a southwest–northeast direction in the lowlands. The lower clay partially restricts the flow to two local systems at the uplands, which converge at the foot of the slope where groundwater primarily flows horizontally before discharging into the wetland. Waters within the upper aquifer are high in NO_3^- and SO_4^{2-} attributable to fertilizer inputs in the croplands (Fig. 2). More than 75% of the samples from the croplands exceeded the maximum recommended values of 45 mg l^{-1} for NO_3^- (WED, 2001), although it showed an important decrease within the water below the forest, where only 23% of the samples presented values above the maximum drinking standards. On the other hand, the near absence of NO_3^- was ubiquitous in the lower valleys and wetland. In these areas, NO_3^- concentrations at some wells increased in autumn and winter, in an inverse relationship with precipitation amounts.

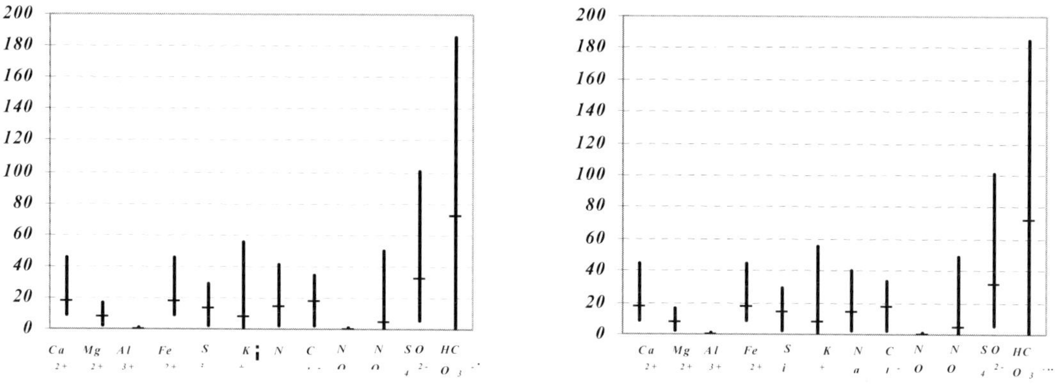

Fig. 2 Mean groundwater quality for the shallow aquifer (left), and the deep aquifer (right). Units, mg l^{-1}.

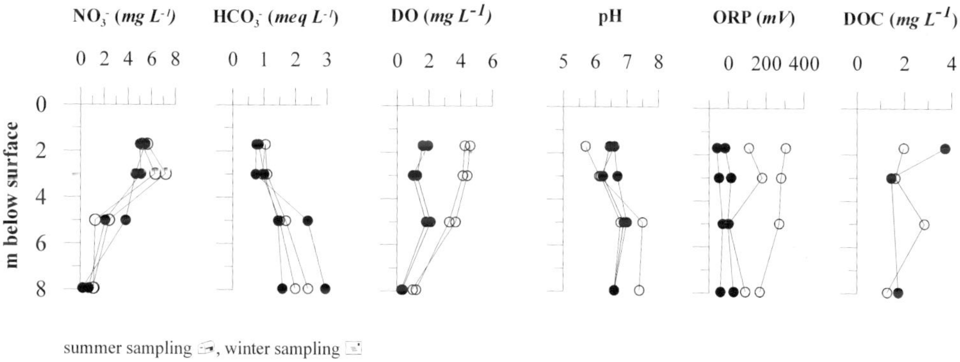

Fig. 3 Average of chemical parameters *vs* depth for the wells near the wetland.

Groundwater in the deep aquifer can be classified as Ca^+-HCO_3^- type. Bicarbonate concentrations averaged 69.3 mg l^{-1}, in contrast to the mean of 40 mg l^{-1} calculated for the shallow unit. The increase in concentrations was gradual with depth, but a more rapid rise in HCO_3^- occurred at the top of the lower aquifer (Fig. 3).

Nitrate concentrations dropped sharply with depth simultaneously with oxygen depletion, indicating the extension of the plume is related to the presence of a redox boundary approximately coincident with the bottom of the clay aquitard in the uplands, and the base of the upper unit in the lowland areas. Dissolved oxygen concentrations also progressively increased at nearly all depths after the end of summer, probably associated with more aerated soils resulting from the decline in the water table levels (Haycock & Pinay, 1993).

Most of the redox measurements were in the range of –50 to 300 mV, with minimum values in the confined section of the lower aquifer and the wetland. In the lower valleys, the oxidation of organic matter would have removed most of the oxygen in the organic-rich layers, causing the redox potential to decline.

Values of pH in shallow groundwater normally fluctuated between six and seven because of the buffering effects from dolomite additions. However, there was a general increase of pH with depth to about 5-m depth as a result of the more elevated HCO_3^- concentrations.

Considering the typical groundwater contains <2 mg l^{-1} of organic carbon (Drever, 1997), measured concentrations are somewhat high, frequently up to 6.5 mg l^{-1}, or even more. Maximums were detected near the wetland. Organic carbon can be supplied to groundwater when it interacts with surface waters or roots of vegetation (Mohamed *et al.*, 2003), or from organic matter within sediments. In the lowland's valleys, DOC would migrate rapidly into groundwater due to the shallow water table. Within the croplands, DOC concentrations remained high even at depths of 10 m below the surface. The systematic decrease with depth suggests the source of OC was not within the deep aquifer itself, but at the surficial sediments. Therefore, it can be concluded that DOC partially survives transport through the unsaturated zone being able to migrate downward into the deep layers.

DISCUSSION

The contaminant plume extended over a well-defined zone near the water table beneath the croplands (Fig. 4), where NO_3^- is thermodynamically stable under the prevailing oxidizing conditions. The highest concentrations were especially related to those fields producing Chinese cabbage, which received the largest fertilizer amounts throughout the area. Domestic wastewater might have exerted some additional influence on the plume generation at the southern edge of the site, but the evidence

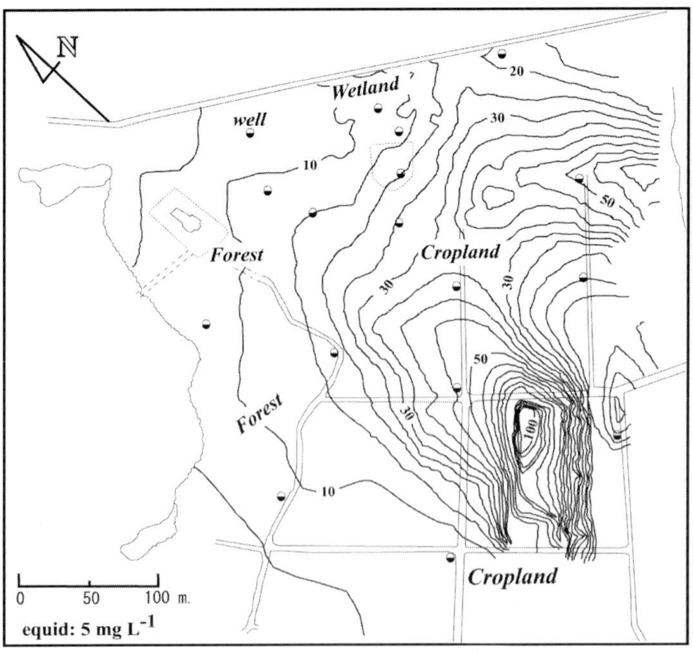

Fig. 4 Distribution of the nitrate plume. Units, mg l^{-1}.

was not conclusive. In contrast, the forested area was generally poor in NO_3^- due to mixing between waters of different chemical signatures. As the plume migrates, waters low in ions recharge the aquifer beneath the woodlands, causing a lens of water relatively depleted in NO_3^- to form near the water table (Kelly, 1997).

Several complex factors take place in the valleys and wetland. Unlike the uplands, NO_3^- in the lowlands is being reduced at shallow depths because of a high water table and a redoxcline close to the surface. Denitrification seems to be the dominant mechanism of NO_3^- reduction. The four requirements for denitrification are (Firestone, 1982): (1) N oxides; (2) the presence of bacteria; (3) electron donors; and (4) restricted O_2 availability. The electron acceptors were supplied by the nitrate-laden groundwater flowing from the agricultural fields, while the presence of bacteria would not be limited in a riparian environment as they are capable of surviving with or without O_2 (Firestone, 1982). In the absence of pyrite, organic carbon supplied by the organic-matter-rich layers would act as the primary electron donor:

$$5CH_2O + 4NO_3^- \rightarrow 2N_2 + 4HCO_3^- + CO_2 + 3H_2O \quad (1)$$

The depletion of NO_3^- was correlated with reductions in DO and the rise in HCO_3^- and pH in groundwater, supporting the hypothesis that denitrification activity is occurring. Furthermore, redox conditions were normally in the range where denitrification can take place, estimated to be below 200 to 300 mV (Kralova *et al.*, 1992). Nitrate concentrations decreased especially when DO values fell below approximately 1.2 mg l^{-1}, close to the threshold of 2 mg l^{-1} found by Cey *et al.* (1999). Thus, although facultative bacteria switch to NO_3^- as an electron acceptor when oxygen supplies become limited (Korom, 1992), denitrification in the studied site would take place before the complete consumption of O_2 in the aquifer. Nitrate retention would be maximum in the first few metres of groundwater flow through the valley corridors and therefore, there would be a significant potential for NO_3^- reduction within the rest of the wetlands, since a large portion of these sites would not be actively denitrifying due to the lack of NO_3^- in groundwater.

Despite of the fact that vegetation uptake might play a role during summer, the primary removal mechanism still seems to be denitrification. There was no vegetation during the dormant season, yet NO_3^- was attenuated, and as soon as vegetation started to grow it was removed by the farmers. On the other hand, removal of vegetation during summer results in an increase of plant and roots litter that in turn might enhance the growth of bacteria and promote higher rates of decomposition. As bacterial activity tends to increase with higher temperatures, greater attenuation rates are therefore expected to occur in the warmer periods (Kelly, 1997).

The near absence of NH_4^+ throughout the catchment indicates that reduction to ammonium was not significant in the attenuation process.

Even though the aquitard beneath the croplands restricts downward solute movement, chemical evidence suggests that conditions at depth might also be favorable for denitrification reactions. As in the lowlands, minimum NO_3^- concentrations coincided with low ORP and DO, and showed an inverse relation with HCO_3^- and pH. The production of HCO_3^-, which would be expected if organic carbon is being oxidized by denitrification, decreases the pH and increases the solubility of carbonate minerals, explaining then the rise in Ca^{2+} concentrations below the nitrate plume (Kelly, 1997). Again, the source of the electron donor would be in the surface soils, as the depth of the water table is not a controlling factor in the transport of OC (Korom, 1992), which still presented values above 2 mg l^{-1} at the top of the deep aquifer. Thus, it is possible that a combination of both the hydrological setting and denitrification activity control the migration of NO_3^- to the lower aquifer.

SUMMARY AND CONCLUSIONS

Although the conclusions are not definitive and much work remains to be done, the present study provided new insights into NO_3^- transport within groundwater in an agricultural area, especially examining the influence of wetlands and stream valleys in the attenuation of pollution.

Nitrate concentrations in groundwater dropped sharply towards the wetland and with depth, suggesting denitrification takes place. Some increases in removal rates in summer are better explained by a rise in denitrification activity rather than vegetation uptake. Attenuation would peak in the first metres of groundwater flow along the valleys, and therefore there would be a significant potential for NO_3^- reduction within the rest of the valleys and wetland.

Nitrate near the water table in the forest was depleted because of mixing between different waters.

The sharp vertical reduction in NO_3^- beneath the croplands might be the result of the effectiveness of the clay layer in preventing its downward migration, but for the portion of the plume able to reach the lower aquifer, denitrification could be the main attenuation mechanism.

REFERENCES

Cey, E. E., Rudolph, D. L., Aravena, R. & Parkin, G. (1999) Role of the riparian zone in controlling the distribution and fate of agricultural nitrogen near a small stream in southern Ontario. *J. Contam. Hydrol.* **37**, 45–67.

Drever, J. I. (1997) *The Geochemistry of Natural Waters: Surface and Groundwater Environments*, 3rd edn. Prentice-Hall, New Jersey, USA.

Firestone, M. K. (1982) Biological denitrification. In: *Nitrogen in Agriculture Soils* (ed. by F. J. Stevenson), 289–326. American Society of Agronomy, Madison, Wisconsin, USA.

Haycock, N. E. & Pinay, G. (1993) Groundwater nitrate dynamics in grass and poplar vegetated riparian buffer strips during the winter. *J. Environ. Qual.* **22**, 273–278.

Kelly, W. R. (1997) Heterogeneities in ground-water geochemistry in a sand aquifer beneath an irrigated field. *J. Hydrol.* **198**, 154–176.

Korom, S. F. (1992) Natural denitrification in the saturated zone: a review. *Water Resour. Res.* **28** (6), 1657–1668.

Kralova, M., Masscheleyn, P. H., Lindau, C. W. & Patrick, W. H. Jr (1992) Production of dinitrogen and nitrous oxide in soil suspensions as affected by redox potential. *Water Air Soil Pollut.* **61**, 37–45.

McDonald, M. G. & Harbaugh, A. W. (1988) A modular three-dimensional finite-difference groundwater flow model. *Techniques of Water Resources Investigations of the US Geological Survey. Book 6.*

Mohamed, A. A. M., Terao, H., Suzuki, R., Babiker, I. S., Ohta, K., Kaori, K. & Kato, K. (2003) Natural denitrification in the Kakamigahara groundwater basin, Gifu prefecture, central Japan. *Sci. Total Environ.* **307**, 191–201.

Postma, D., Boesen, C., Kristiansen, H. & Larsen, F. (1991) Nitrate reduction in an unconfined sandy aquifer: water chemistry, reduction processes, and geochemical modeling. *Water Resour. Res.* **27** (8), 2027–2045.

Sugawara, Y (2004) Three-dimensional nitrate dynamics in groundwater at the edge of upland. MSc Thesis, University of Tsukuba, Ibaraki, Japan (in Japanese).

Water Environmental Department (2001) *Water Environment Management in Japan*. Environmental Management Bureau, Ministry of the Environment, Tokyo, Japan.

Dynamics and Biogeochemistry of River Corridors and Wetlands (Proceedings of symposium S4 held during the Seventh IAHS Scientific Assembly at Foz do Iguaçu, Brazil, April 2005). IAHS Publ. 294, 2005.

Integrated river basin modelling including wetlands and riparian zones in the German Elbe River basin

FRED F. HATTERMANN, VALENTINA KRYSANOVA & ANJA HABECK

Potsdam Institute for Climate Impact Research (PIK), Potsdam, Germany
hattermann@pik-potsdam.de

Abstract Riparian wetlands are hydrologically connected to both the surrounding upstream catchment area and the river, and represent an interface between them. They intercept surface and subsurface runoff with soluble nutrients and sediments from the upland areas, and therefore function as buffers for the river network. Due to their connection to groundwater and the supply of nutrients from upland areas, riparian wetlands have excellent conditions for vegetation development during the whole growth season. As a result, riparian wetlands are highly effective in the reduction of diffuse source pollution and sediment loads to rivers and streams. Therefore integrating wetlands and riparian zones in eco-hydrological river basin modelling is very challenging. But the correct representation of all these processes in the modelling framework with their different characteristic spatial and temporal scales is not a trivial task and includes a lot of inherent uncertainty. Additional problems are the identification of riparian zones based on regionally available data, and verification of the results.

Key words groundwater dynamics; nutrient retention; riparian zones; water quality; wetlands

INTRODUCTION

The Water Framework Directive of the European Commission demands the bringing of water bodies in Europe to "a good ecological status" (EC 2000). Many efforts and improvements have been done, mainly in the implementation of wastewater treatment plants. But these measures only help to improve the water quality of point sources, whereas the main origin of some important contaminants are diffuse sources like atmospheric deposition and fertilization of crop land. Here, riparian zones and wetlands play an important role in the control of the water quality of surface water systems (Dall'O' *et al.*, 2001).

The paper presents an integrated catchment model with which it is possible to analyse the processes in wetlands and riparian zones in meso- to macroscale river basins, the scale relevant for water management planning and for the implementation of the Water Framework Directive. A simple but comprehensive mechanistic wetland module was developed and coupled with the eco-hydrological model SWIM (Soil and Water Integrated Model; Krysanova *et al.*, 1998), which integrates hydrological processes, vegetation, erosion and nutrient dynamics at the watershed scale. The reliability of the model results was tested under well defined boundary conditions by comparing the results with those from a two-dimensional (2-D) numeric groundwater model under steady-state and transient conditions (Hattermann *et al.*, 2004b) as well as with observed data of a meso-scale basin, using contour maps of the long-term mean water table, observed groundwater level data, and observed river discharge and nutrient concentrations.

The results of the study show that riparian zones and wetlands have a high potential to reduce the nutrient transport into surface water systems, although the uncertainty of the modelling results is very high. Their impact is so large because they are at the interface between catchment and river systems, where the greater part of the nutrients in the catchment originally applied as fertilizers or mineralized from plant residues is already degraded. Restoration and management of wetlands is therefore of high priority for the control of non point source contamination of surface waters.

MATERIAL AND METHODS

The model

SWIM A three-level scheme of spatial disaggregation from basin to sub-basins and to hydrotopes is used in the model. A hydrotope is a set of elementary units in the sub-basin, which have the same geographical features, e.g. land use, soil type, and average water table depth. Therefore it can be assumed that they behave in a hydrologically uniform way (Krysanova *et al.*, 2000). Water fluxes, plant growth and nitrogen dynamics are calculated for every hydrotope, where up to 60 vertical soil layers can be considered. The outputs from the hydrotopes are aggregated at the sub-basin scale. Mean resistance time and potential retention of water and nutrient fluxes are calculated using spatial features of the hydrotopes like distance to next river, gradient of the groundwater table and permeability of the aquifer. The approach allows us to consider and investigate the spatial pattern of land-use and land-use changes. The lateral fluxes are routed over the river network, taking transmission losses into account. Plant dynamics are simulated using a simplified EPIC approach (Williams *et al.*, 1984). A full description of the model can be found in Krysanova *et al.* (1998, 2000). An extensive hydrological validation of the model in the Elbe basin including sensitivity and uncertainty analyses is described in Hattermann *et al.* (2004a).

The wetland module It is important for the investigation of meso- to macroscale river basins to apply methods which are physically sound but simple enough to save computation time and data demand (Arnold *et al.*, 1993). The wetland module described here consists of two parts: one part describes the groundwater fluxes and water table dynamics, where the time scale is of days or weeks. The second part describes the nutrient fluxes and degradation, where the time scale is much larger (years and decades, sometimes centuries, because of the mean residence time of the groundwater).

A good reproduction of the groundwater dynamics is important for the hydrological processes and nutrient fluxes in wetlands. Smedema & Rycroft (1983) derived a linear storage equation following the Dupuit-Forchheimer assumptions to predict the non-steady-state response of groundwater flow to periodic recharge from Hooghoudt's steady-state formula, assuming that the time-derivative of return flow q in mm day^{-1}

(the groundwater discharge) at time step t is linearly related to the rate of change in water table height h in m (water table above drainage base), whereby only headlosses in horizontal direction are considered:

$$\frac{dq}{dt} = \frac{8*T}{L^2} * \frac{dh}{dt} \tag{1}$$

where T is the transmissivity in m^2 day^{-1} and L the slope length in m. If the groundwater body is recharged by deep soil percolation or another source (Rc in mm day^{-1}) and is depleted by drain discharge (q), it follows that the water table will rise when $Rc - q > 0$ and fall when $Rc - q < 0$. The water table fluctuations may be described as:

$$\frac{dh}{dt} = \frac{(Rc - q)}{C*S} \tag{2}$$

S is again the specific yield. It follows that by assuming that the retention constant $C = 0.8$ (Smedema & Rycroft, 1983):

$$\frac{dq}{dt} = \frac{10*T}{S*L^2} * (Rc - q) = \alpha * (Rc - q) \tag{3}$$

so that the change in drain discharge dq/dt is proportional to the excess recharge $Rc - q$, with α being the proportionality factor (reaction factor, see equation (6)). Equation (2) can be transformed to gain the equation for return flow:

$$q_t = q_{t-1} * \exp(-\alpha * \Delta t) + Rc_{\Delta t} * (1 - \exp(-\alpha * \Delta t) \tag{4}$$

Using the linear relationship between q and h (equation (1)), we get:

$$h_t = h_{t-1} * \exp(-\alpha * \Delta t)) + \frac{Rc_{\Delta t}}{0.8 * S * \alpha} * (1 - \exp(-\alpha * \Delta t) \tag{5}$$

The equations are scale independent and the spatial unit for which h and q are calculated can be either the hydrotope or the sub-basin.

The factor α is a function of the transmissivity T and the slope length L:

$$\alpha = \frac{10*T}{S*L^2} \tag{6}$$

Therefore, the reaction factor has a physical meaning and can be estimated directly by using observations of the groundwater head h. This was done using an automatic calibration algorithm by adjusting T and S in physically sound limits.

While it is possible to describe water table dynamics using the mean reaction time, the time scales which have to be considered for the simulation of nutrient retention are much larger (years and decades), because the actual residence time is the crucial value which determines the intensity of degradation. According to Wendland *et al.* (1993), the degradation of nitrate N (kg ha^{-1}) can be approximated by a linear decay equation, where λ is a function of temperature and available oxygen. The full retention of a landscape is then a function of mean residence time and degradation:

$$\frac{dN_{t,out}}{dt} = N_{t,in} \frac{1}{1 + K\lambda} (1 - e^{-(\frac{1}{K} + \lambda)t}) + N_{t-1,out} e^{-(\frac{1}{K} + \lambda)t} \tag{7}$$

where *K* in days is the mean residence time. The mean residence time of water in the sub-basin to flow from a specific hydrotope to the next river is calculated for each hydrotope using GIS and digital maps of the groundwater table and the geo-hydrology. Since SWIM distinguishes between surface flow, interflow and base flow, each having different retention characteristics (residence time, oxygen content), there has to be one equation for each of the fluxes.

Plant uptake of water and nutrients from groundwater is only possible in times when the plant roots have excess to it and if the plant demand cannot be satisfied by soil water and nutrient recourses. A resistance function controls the ability of plant roots for water and nutrient uptake from groundwater.

The basin

The northern lowland part of the German Elbe basin, where the model was tested in the Nuthe catchment (1938.0 km^2, see Fig. 1), is climatically one of the driest regions in Germany, with mean annual precipitation of about 600 mm year^{-1}. Hence, water availability during the summer season is the limiting factor for plant growth. The lowland is formed by mostly sandy glacial sediments and drained by slowly flowing streams with broad river valleys. The upper sites with deep water tables are covered by sandy, highly permeable soils and mostly pine forests or by arable land on ground moraine with till soils that tend to have layers with lower water permeability.

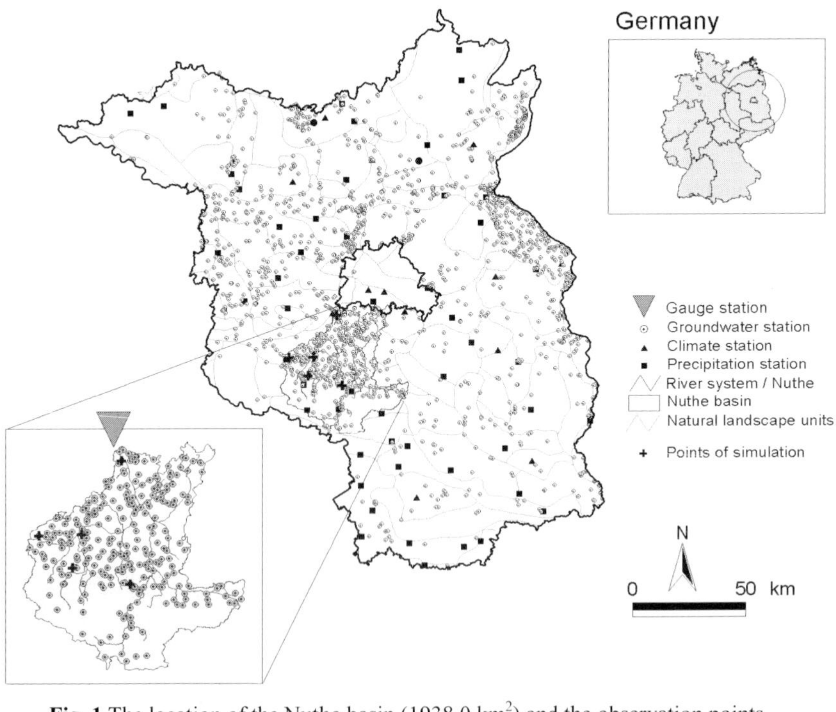

Fig. 1 The location of the Nuthe basin (1938.0 km^2) and the observation points.

Valleys are covered by loamy alluvial soils with grassland and riparian forests, where the groundwater is very shallow, and arable land elsewhere.

RESULTS AND DISCUSSION

Groundwater and river flow dynamics

First, the simulated mean annual water table depth of all sub-basins in the Nuthe basins were calibrated automatically using the permeability in a physically sound range (8–60 m day^{-1}). The mean simulated amplitude was too high and had to be smoothed by a moderate increase in the value of specific yield (2.5–30%, as taken from the geohydrological map). The Mean Absolute Error of the long-term observed against the simulated water table in all sub-basins was 0.026 m.

The observation wells were selected in order to represent a cross section through the basin from the lowlands in the north to the hilly area in the southwest. Well 1 is located next to the outlet of the Nuthe River catchment. The curves show a good fit, especially for the early 1980s. The rise of the groundwater level in 1987 and 1988 is slightly overestimated by the model in sub-basins 2, 4 and 5. As explained in section 1, the natural flow regime in the Nuthe basin is influenced by stream flow control (weir and reservoir management), and especially in the lowland areas, the water level is controlled by land drainage.

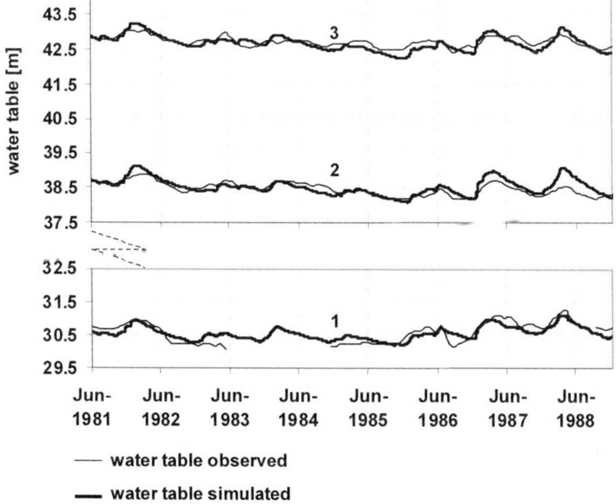

Fig. 2 Comparison of observed and simulated groundwater table for five locations in the Nuthe basin (Hattermann *et al.*, 2004b).

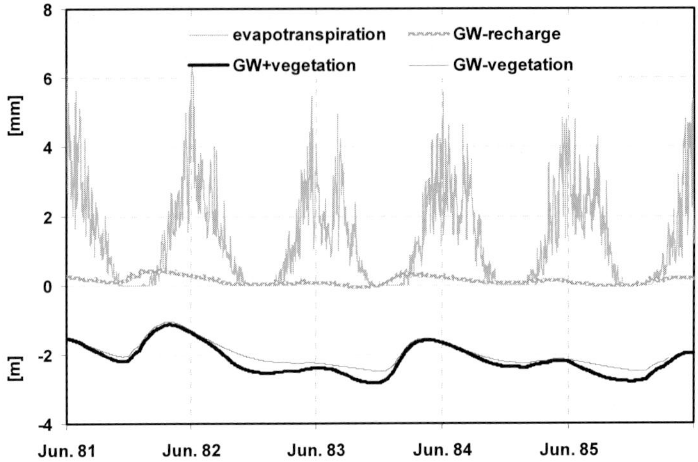

Fig. 3 Comparison of simulated and observed groundwater table with (GW + vegetation) and without plant water uptake from groundwater (GW – vegetation).

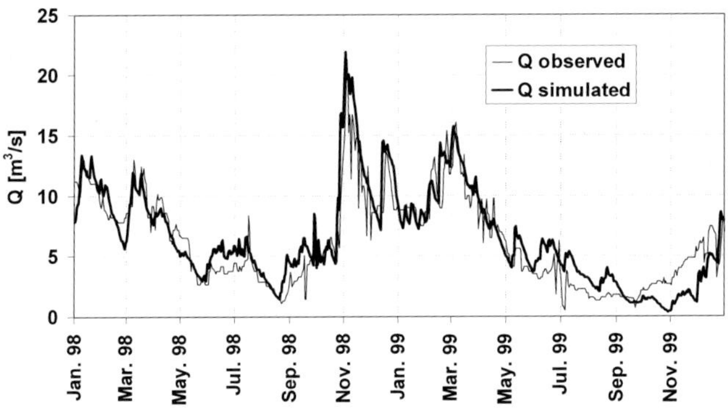

Fig. 4 Comparison of daily river flow observed and simulated (gauge Babelsberg).

The simulated groundwater hydrographs are very similar, whereas the observations show more differences. The higher variability in the observed water levels is the result of small-scale heterogeneities in the aquifer and of local precipitation events which are missing in the observed records.

Figure 3 illustrates the impact of plant water uptake on the simulated water table. While the groundwater tables simulated with and without plant water uptake converge during the winter term, they separate during the vegetation period, where the plant uptake leads to a decline of the groundwater table. Evapotranspiration and recharge in Fig. 3 are calculated with plant water uptake from groundwater.

The mean long term difference between the observed and simulated river discharge at the basin outlet is 3.0% for the calibration period 1981–1988, indicating that the water balance is correctly calculated by SWIM. The daily Nash & Sutcliffe efficiency is 0.7 (only 0.54 for the validation period 1989–2000). The hydraulic regime of the Nuthe basin is strongly influenced by water management regulations like drainage systems and weir plants, so that it is difficult to reproduce the hydrograph with higher accuracy. The summer discharge is in some years overestimated by the model (see Fig. 4).

This can be explained by water abstraction and regulation measures, when a minimum river flow is provided by reservoir management in dry summer periods. It is worth mentioning that the efficiency was notably higher for other meso- and macro-scale sub-basins of the Elbe located in hilly and mountainous areas (Hattermann *et al.*, 2004a). Without additional plant water uptake from groundwater, the total evapotranspiration would be 25% lower, leading to an increase in river discharge of about 50%.

Nitrate concentrations

The nitrate concentration in the Nuthe River during the 1980s was strongly influenced by point sources (irrigation of wastewaters in very small areas, municipal wastewaters, even direct discharge of liquid manure into surface waters), where the records are vague and incomplete, so that the comparison in Fig. 5 is done for a time period in the 1990s, where the impact of point sources is very limited because of the implementation of wastewater treatment plants in the basin. Diffuse sources in this study are fertilizer applications (about 135 kg ha^{-1} for winter wheat, a smaller application in late autumn, a larger in spring), atmospheric deposition (about 25 kg ha^{-1}), and plant residues after harvest and autumn.

The comparison shows that the periodicity and amplitude of the observed values is mostly well reproduced by SWIM, although the difference between observed and simulated values is large especially at the end of the year. The reason is that the diffuse sources for nitrate contamination (in particular fertilization) are not very well known, because information about crop rotation schemes and fertilization regimes are not available at the regional scale.

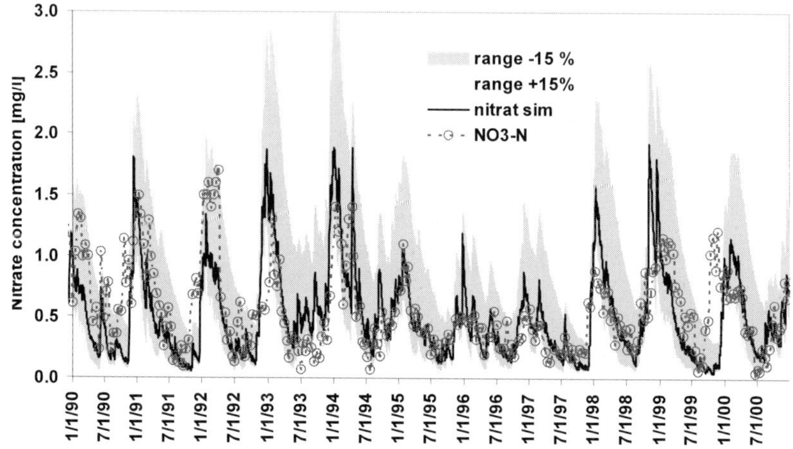

Fig. 5 Simulated and observed nitrate concentrations in the Nuthe River with the range of uncertainty because of uncertain geo-hydrological information.

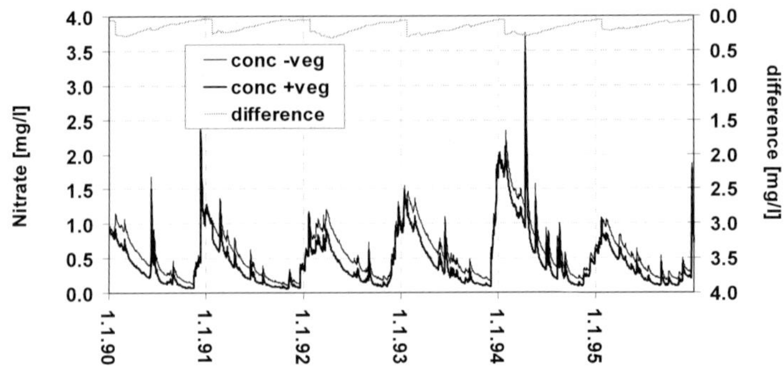

Fig. 6 Comparison of simulated and observed nitrate concentration with and without plant water uptake from groundwater.

The mean residence time of groundwater is 41 years, with a maximum of approximately 400 years. The values are in good agreement with Landesumwelamt (2002), who estimated the nutrient loads and retention in the lowland catchments of the Elbe basin. The range of uncertainty produced by the uncertain geohydrological information is also included in Fig. 5 by assuming a 15% longer or shorter mean residence time and half life time.

Figure 6 illustrates the impact of plant uptake of nitrate in riparian zones and wetlands. As also shown for the impacts of plants on the water level in Fig. 3, the differences are the highest during the summer season, when plant demand is high and can therefore not be satisfied by the soil water concentrations. The difference becomes smaller during the late summer, because the total amount of available nutrients in soils and hence also the leaching of nutrients has its minimum. Figure 7 shows a map of the additional plant nitrate uptake from groundwater in kg ha^{-1}. The values are not so large

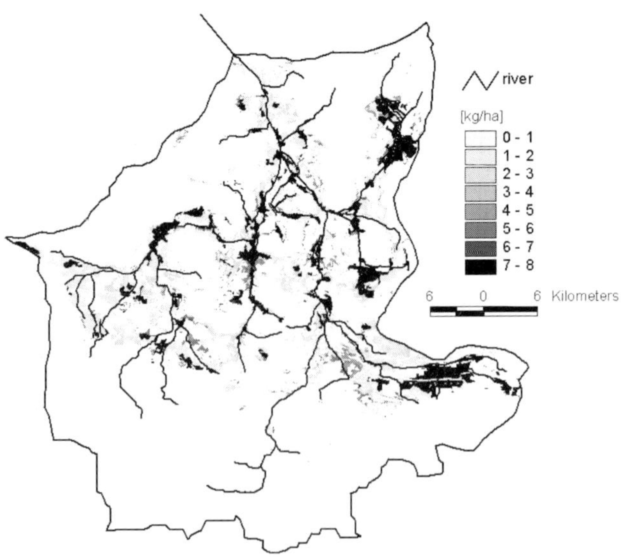

Fig. 7 Additional nitrate uptake by plants in riparian zones and wetlands.

in comparison with the total plant uptake (up to 150 kg ha^{-1}). The additional uptake is only about 1% of the total uptake, but this leads to a retention of about 24% of the total river load. The reason is that the additional uptake happens in an area next to the surface water bodies, where the largest part of the nitrate which was originally applied by fertilizers, mineralized from plant residues and decomposed from the atmosphere is already degraded.

CONCLUSIONS

The simulation results indicate that relatively small parts of the total catchment area have a relatively high impact on the water and nutrient balance in the catchment (additional evapotranspiration of about 25%, additional nitrate uptake of about 1%, leading to a decrease in river discharge of about 50% and to a decrease in annual river nitrate load of about 24%), although the uncertainty of the results is high. Riparian zones and wetlands are buffer systems able to reduce contamination of surface waters, as long as the vegetation has access to groundwater. On the other hand, restoration of wetlands will lead to increased water losses by evapotranspiration, crucial in a region where river discharge during the summer season is only possible by water regulation through dams and weirs, and where a trend to lower annual precipitation has been observed during the last decades. It follows that water managers have to find a sensitive balance between water quality and water quantity aspects in the planning process.

REFERENCES

Arnold, J. G., Allen, P. M. & Bernhardt, A. (1993) A comprehensive surface-groundwater flow model. *J. Hydrol.* **142**, 47–69.

Dall'O', M., Kluge, W. & Bartels, F. (2001) FEUWAnet: a multi-box water level and lateral exchange model for riparian wetlands. *J. Hydrol.* **250**, 40–62.

EC (2000) Establishing a framework for community action in the field of water policy. Directive 2000/60/EC of the European Parliament and of the Council of 23 October 2000. *Official Journal of the European Communities*, Brussels, Belgium.

Hattermann, F. F., Wattenbach, M., Krysanova, V. & Wechsung, F. (2004a) Macroscale validation of the eco-hydrological model SWIM for hydrological processes in the Elbe basin with uncertainty analysis. In: *Regional Assessment of Climate and Management Impacts Using the SWAT Hydrological Model* (ed. by N. Fohrer & J. Arnold). *Hydrological Processes* (in press).

Hattermann, F. F., Krysanova, V., Wechsung, F., Wattenbach, M. (2004b) Integrating groundwater dynamics in regional hydrological modelling. *Environ. Modelling Software* **19**(11), 1039–1051.

Krysanova, V., Becker, A. & Müller-Wohlfeil, D.-I. (1998) Development and test of a spatially distributed hydrological / water quality model for mesoscale watersheds. *Ecol. Modelling* **106**, 261–289.

Krysanova, V., Wechsung, F., Arnold, J., Srinivasan, R. & Williams, J. (2000) SWIM (Soil and Water Integrated Model) User Manual. *PIK Report no. 69.* PIK, Potsdam, Germany.

Landesumweltamt Brandenburg (2002). Stoffeinträge in die Gewässer des Landes Brandenburg. Band 68. Potsdam, Germany.

Smedema, L. K. & Rycroft, D. W. (1983) *Land Drainage—Planning and Design of Agricultural Drainage Systems.* Cornell University Press, Ithaca, New York, USA.

Wendtland, F., Albert, H., Bach, M. & Schmidt, R. (1993) *Atlas zum Nitratstrom in der Bundesrepublik Deutschland.* Springer Verlag, Berlin, Germany.

Williams, J. R., Renard, K. G. & Dyke, P. T. (1984) EPIC—a new model for assessing erosion's effect on soil productivity. *J. Soil Water Conserv.* **38**, 381–383.

Relation of corridor vegetation density and stream temperature

KOSHI YOSHIDA[1], HIROAKI SOMURA[2], KATSUHIRO HIGUCHI[2], OSAMU TODA[3] & HAJIME TANJI[2]

1 *Japan Science and Technology Agency, 2-1-6 Kannondai, Tsukuba, Ibaraki 305-8609, Japan* koshiy@nkk.affrc.go.jp

2 *National Institute for Rural Engineering, 2-1-6 Kannondai, Tsukuba, Ibaraki 305-8609, Japan*

3 *Graduate School of University Tokyo, 1-1-1 Yayoi, Bunkyo-ku, Tokyo 113-8657, Japan*

Abstract The decline of river corridor vegetation density strongly affects the river thermal environment, which also has a large impact on the ecological systems. In this study, a stream temperature analysis model related to river corridor vegetation density was developed and applied to the Nam Song River in Laos. Meteorological data (air temperature, relative humidity, wind speed) were used, and the sink or source terms of the heat transfer equation were estimated from those data. Land cover GIS data was used for vegetation density analysis along the main river. The results of simulation with several levels of vegetation density indicate that river water temperature is more sensitive to river discharge change in the case of low vegetation density.

Key words heat transfer equation; Laos; Mekong River; river corridor conservation; river thermal environment; solar radiation estimation; vegetation density

INTRODUCTION

Recently, the multiple functions of river corridors have attracted the attention of biologists and environmental river engineers. In a river with luxuriant corridor vegetation, direct solar radiation is reduced by the crown of trees, thus the stream temperature is kept low, and excessive growth of algae is prevented even during low discharge periods. As the stream temperature is kept low, any rapid increase of immigrant plants, which seize the habitats of conventional plants, is prevented, and as a result, the biodiversity of the habitats is retained (Swift & Messer, 1971).

Temperature is an important parameter in streamwater quality. Human activity raises or lowers natural streamwater temperatures due to impoundments, industrial use, irrigation and global warming (LeBlanc *et al.*, 1979; Bartholow, 1989). Water temperature affects nearly every physical property related to water quality management and, unfortunately, a change of water temperature is often undesirable. Therefore, control of streamwater temperature becomes a prerequisite for maximum utilization of water resources (Yoshida *et al.*, 2004).

The Mekong River, a major international river flowing thorough six countries, had had relatively pristine conditions in previous centuries. However, the habitats of river biota and fish are now threatened by serious changes to river environments resulting from the rapid expansion of human activities. In Laos especially, located in a mountainous area, the river corridors are being destroyed by farmland development or fire agriculture. The decline of river corridor vegetation density strongly affects the river's thermal environment, which has a large impact on the ecological system (Stanford & Ward, 1988).

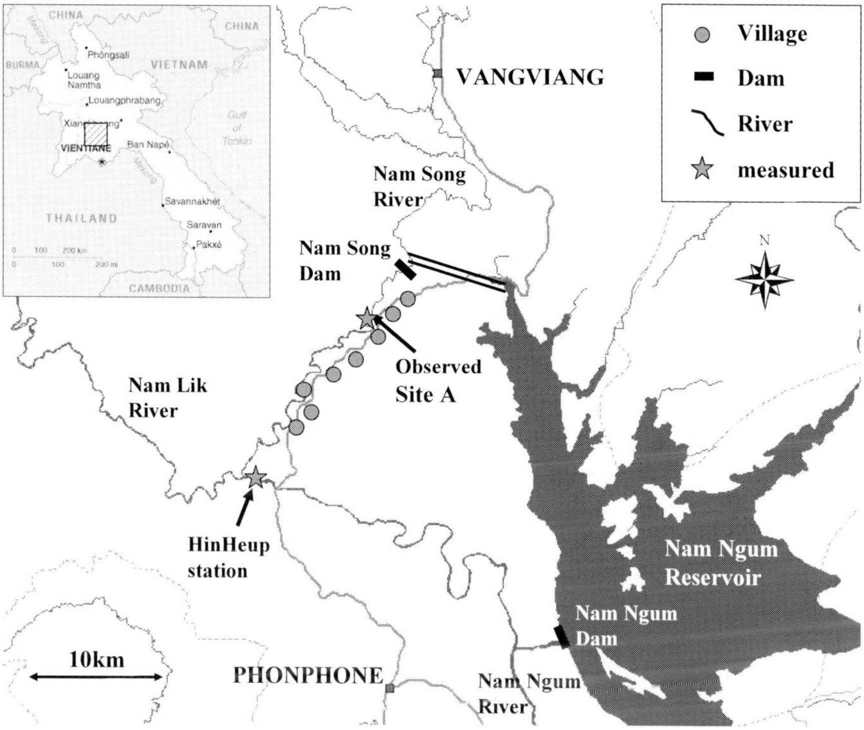

Fig.1 Location of study area.

In this study, a model that relates stream temperature with river corridor vegetation density was developed. The Nam Song River, which is a tributary of the Nam Ngum River flowing through the Vientiane Metropolis, was selected for field study. Meteorological data (air temperature, relative humidity, wind speed) were used, and the sink or source terms of the heat transfer equation were estimated from that data. Land cover GIS data was used for vegetation density analysis along the main river. Simulations were made with several cases of vegetation density, showing that river water temperature is more sensitive to river discharge change in the case of low vegetation density.

STUDY AREA

The Nam Song River, having a 1800 km^2 catchment area and 100 km length, is a tributary of the Nam Ngum River (Fig. 1). The Nam Song Diversion Dam (1500 km^2 catchment area) was constructed in 1996 to divert water from the Nam Song basin to the Nam Ngum reservoir. Downstream of the Nam Song Dam, there are eight villages consisting of approximately 4000 people along the riverbanks. In the dry season, the water release discharge from the Nam Song Dam decreased from 10.0 m^3 s^{-1} to a minimum of only 2.0 m^3 s^{-1} (as shown in Fig. 2, estimated by Yoshida *et al.*, 2004); the

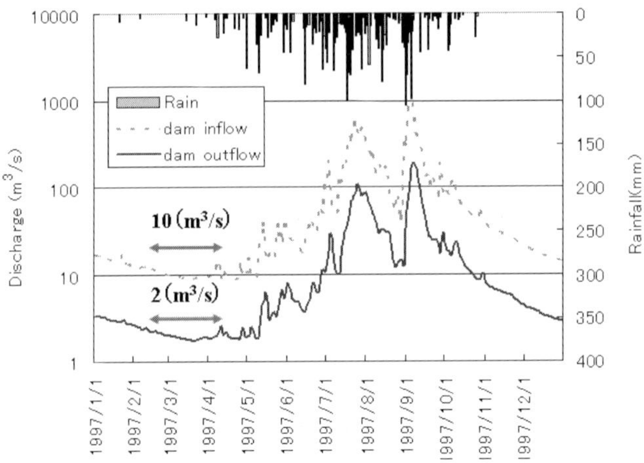

Fig. 2 Hydrograph at the Nam Song diversion dam (1997).

dry season discharge decrease made a large impact on the fisheries and domestic water quality. The decline of the fish catch was a serious loss to local villages, and the river downstream of the dam now has a very reduced number of species. The local people had utilized the forest resources along the river before dam construction. However, many kinds of crops are now transplanted along the riverbank to improve their living as an alternative to fishing. The reduction of the corridor vegetation density also accelerated the environmental decline of the Nam Song River (Roel & Sean, 2001). In this study area, the actual stream temperature was observed 4 km downstream from the Nam Song Dam and daily meteorological data at the Hinheup station was available.

ESTIMATION OF SOLAR RADIATION

To predict the stream temperature, the heat gained or lost from a parcel of water should be calculated as it passes through a stream segment. This is accomplished by simulating the various heat flux processes that determine that temperature change. These physical processes include convection, conduction, evaporation, as well as long wave radiation (heat to or from the air), direct solar radiation (short wave), and radiation back from the water. So first, solar radiation, which is a basal parameter, should be calculated. To calculate solar radiation, the radiation at the outer edge of the Earth's atmosphere was calculated.

In this study, the FAO (Food and Agriculture Organization of the United Nations) method was employed to estimate hourly solar radiation from meteorological data (air temperature, relative humidity, wind speed). The extraterrestrial radiation, R_a, for each day and each hour of the year and for different latitudes can be estimated from the solar constant as follows:

$$R_a = \frac{12(60)}{\pi} G_{sc} [(\omega_2 - \omega_1)\sin(\varphi)\sin(\delta) + \cos(\varphi)\cos(\delta)(\sin(\omega_2) - \sin(\omega_1))] \quad (1)$$

$$dr = 1 + 0.33\cos\left(\frac{2\pi}{365}J\right) \tag{2}$$

$$\delta = 0.409\sin\left(\frac{2\pi}{365}J - 1.39\right) \tag{3}$$

$$\omega_1 = \omega - \frac{\pi\Delta t}{24} \tag{4}$$

$$\omega_2 = \omega + \frac{\pi\Delta t}{24} \tag{5}$$

$$\omega_s = \arccos[-\tan(\varphi)\tan(\delta)] \tag{6}$$

where R_a is hourly extraterrestrial radiation (MJ m^{-2} h^{-1}), G_{sc} is the solar constant = 0.082 (MJ m^{-2} min^{-1}), dr is the inverse relative distance Earth–Sun, φ is latitude (rad), δ is solar declination (rad), J is the number of the day in the year, ω_1 is solar time angle at beginning of period, ω_2 is solar time angle at end of period, ω is solar time angle at midpoint of hourly period (rad), ω_s is the sunset hour angle (rad), Δt is calculation time step (h).

The solar time angle at the midpoint of the period is:

$$\omega = \frac{\pi}{12}[(t_s + 0.06667(L_z - L_m) + S_c) - 12] \tag{7}$$

$$S_c = 0.1645\sin(2f) - 0.1255\cos(f) - 0.025\sin(f) \tag{8}$$

$$f = \frac{2\pi(J - 81)}{364} \tag{9}$$

where t_s is standard clock time at the period's midpoint (e.g. 7.5 for between 07:00 h and 08:00 h), L_z is the longitude of the centre of the local time zone, degrees west of Greenwich (e.g. 255° for Bangkok), L_m is longitude of the measurement site, S_c is seasonal correction for solar time (hour). Of course, $\omega < -\omega_s$ *or* $\omega > \omega_s$ indicates that the sun is below the horizon and R_a is zero.

The solar radiation R_s can be calculated with the Angstrom formula, which relates solar radiation to extraterrestrial radiation and relative sunshine duration:

$$R_s = \left(a_s + b_s\frac{n}{N}\right)R_a \tag{10}$$

$$N = \frac{24}{\pi}\omega_s \tag{11}$$

where R_s is solar or shortwave radiation (MJ m^{-2} h^{-1}), n is actual duration of sunshine (h), N is maximum possible duration of sunshine or daylight hours (h), a_s and b_s are the Angstrom coefficients. Depending on atmospheric conditions (humidity, dust) and solar declination (latitude and month), the Angstrom coefficients will vary. In this study, the FAO recommended values, $a_s = 0.25$ and $b_s = 0.5$, were used.

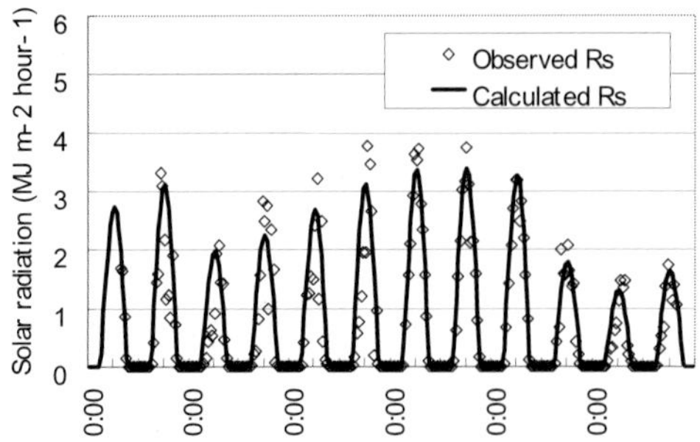

Fig. 3 Comparison of observed and calculated solar radiation (4–14 August 2004).

Figure 3 shows the calculated solar radiation and that observed from 4 to 16 August 2004 at Vientiane Meteorology station (102.5°E, 18°N). There were relatively large margins of error in some radiation peaks; however, overall the daily trend had good agreement with observed values. In this study, to express the solar radiation reduction due to cloudy skies, relative sunshine duration (*n*/*N*) was used in equation (10) during all daylight hours, because only daily sunshine data was available. Using the hourly cloud cover data would have improved the accuracy of estimation, but the data were unavailable.

ESTIMATION OF THE SHADE PARAMETER OF THE RIVER CORRIOR VEGETATION

To evaluate the stream temperature regulation function of the river corridor vegetation, the vegetation shade parameter V_{shade} should be estimated. At first, characteristics of the corridor vegetation such as height V_h, density V_d and vegetation offset V_o from the stream were analysed from GIS land cover data (1-km resolution), provided by the Global Mapping Project (Fig. 4) and field survey data. The Global Map land cover GIS data contains the information on vegetation height and density (e.g. evergreen: height >2 m, density >60%). Table 1 shows the estimated present condition of the river corridor vegetation, which was evaluated as an average value along the main stream from the GIS definition and was modified by field survey). The vegetation density was the same, 60%, for both the right and left banks. However, the right bank vegetation which was better sunlit in the morning was more luxuriant.

From the vegetation characteristics, the local sunrise/sunset time and hourly sun angle, the vegetation shade ratio that considers the sunlight filtered out by the vegetation was calculated (Fig. 5). The watershed-scale radiation shade model was proposed by Chen (1998) and considered complex vegetation shading and topographic effects. In this study, a simplified model was proposed to enable application at a large-scale resolution (1-km mesh). In the 1-km resolution analysis, topographic effects were neglected because 1 km is too much longer than the vegetation offset distance.

The shadow length of the tree vegetation was estimated as follows:

$$S_{length,L} = \left| V_{hL} / \tan(\omega) \right| \quad (-\omega_s < \omega < 0) \tag{12}$$

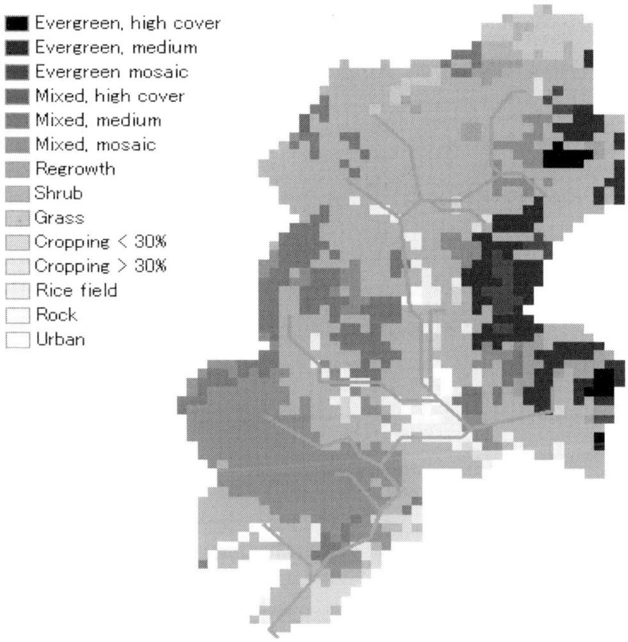

Fig. 4 GIS land cover data of Nam Song watershed.

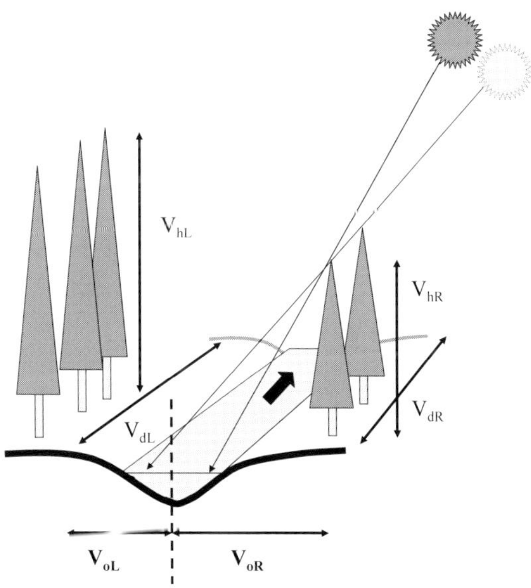

Fig. 5 Sketch of the shade ratio calculation from sun angle and vegetation data.

Table 1 Characteristics of river corridor vegetation.

	V_h(m)	V_d(%)	V_o(m)
Left bank	7.0	60.0	10.0
Right bank	10.0	60.0	7.0

V_h : vegetation height; V_d : vegetation density; V_o: vegetation offset from the stream.

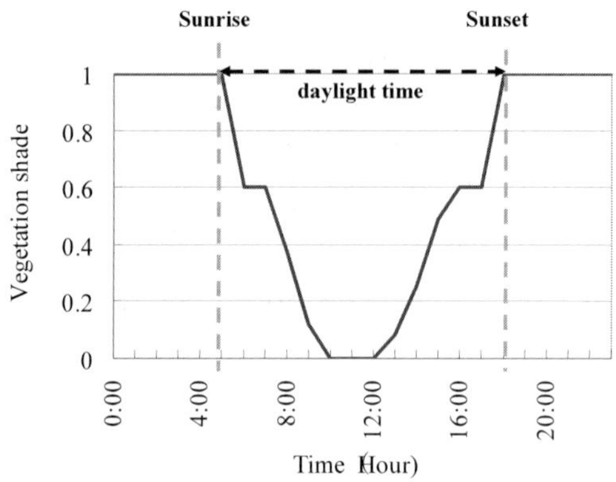

Fig. 6 Daily change of the vegetation shade ratio (calculated on 31 March 2004).

$$S_{length,R} = |V_{hR} / \tan(\omega)| \quad (0 \le \omega < \omega_s) \tag{13}$$

where S_{length} is the shadow length of the riparian tree, V_{hL} and V_{hR} are vegetation height of left and right side bank.

Shined (sunlit) water surface width was calculated by:

$$W_{shined,L} = \max(0, V_{0L} + W/2 - S_{length,L}) \quad (-\omega_s < \omega < 0) \tag{14}$$

$$W_{shined,R} = \max(0, V_{0R} + W/2 - S_{length,R}) \quad (0 < \omega < \omega_s) \tag{15}$$

where W_{shined} is shined water surface width, V_{oL} is vegetation offset distance, W is water surface width.

The vegetation shade ratio was calculated as:

$$\begin{aligned} V_{shade} &= V_{dL}(W - W_{shined,L})/W \quad (-\omega_s < \omega < 0) \\ &= V_{dR}(W - W_{shined,R})/W \quad (-\omega_s < \omega < 0) \end{aligned} \tag{16}$$

where V_{shade} is vegetation shade ratio, and V_d is vegetation density along the stream.

Figure 6 shows the analysed result of the daily change of shade ratio. Vegetation shaded time was longer in the afternoon due to the luxuriant right bank vegetation and the value was zero around noon.

STREAM TEMPERATURE ANALYSIS

In this study, the one-dimensional heat advection–dispersion equation including the heat exchange with the atmosphere was used. When applied to an open channel of variable cross section and flow rate, the heat transport equation takes the form:

$$A\frac{\partial T}{\partial t} + \frac{\partial (QT)}{\partial x} = \frac{\partial}{\partial x}\left(AD_L\frac{\partial T}{\partial x}\right) + \frac{WS}{\rho Cp} + \frac{G}{\rho Cp} \tag{17}$$

where T is water temperature, x is distance, t is time, D_L is a dispersion coefficient in x direction, S is a source or sink term which includes heat transfer with the surrounding environment, A is cross-sectional area, W is surface width, Q is flow rate, G is the lateral heat advection term by lateral inflow. For numerical testing, dispersion values (D_L) of 20–200 $m^2\ s^{-1}$ were applied assuming several river conditions (Bashar & Stefan, 1993), but increases of D_L did not change the model simulation results because of the relatively high velocity of flow and the low longitudinal temperature gradient along the modelled stream, indicating that the system was advection dominant. Therefore D_L was set equal to zero in the model.

Equation (17) can be used to predict water temperatures in steady flow streams that are well mixed and have no significant transverse temperature gradients; the main variation is in the flow direction. Stream cross-sectional area A and width W are a function of stream flow rate Q, ρ is the density of water, and Cp is the specific heat of water.

The source or sink term S expresses the heat transfer rate with the surrounding environment:

$$S = S_a + S_b \tag{18}$$

with

$$S_a = H_s + H_l + H_e + H_c \tag{19}$$

where S_a is the net heat exchange rate between the water and the air, S_b is the net heat exchange between the water and the streambed, H_s is the net short-wave (solar) radiation, H_l is the net long-wave radiation, H_e is the evaporative heat transfer, H_c is the convective heat transfer.

H_s is the net solar radiation. Solar radiation is a measured weather parameter, and net shortwave radiation is expressed as the difference between measured incoming and reflected radiation:

$$H_s = (1-\alpha)Rs(1-V_{shade}) \tag{20}$$

where α (= 0.06 was employed) is the reflection parameter at the water surface, R_s is solar radiation and V_{shade} is vegetation shade ratio estimated by corridor vegetation analysis.

H_l is the net long-wave radiation and can be expressed as:

$$H_l = \sigma\left(\varepsilon_a T_{K,a}^4 - \varepsilon_w T_{K,s}^4\right) \tag{21}$$

where σ is the Stefan-Boltzmann constant, $T_{K,a}$ is air temperature (degrees Kelvin), $T_{K,s}$ is the water surface temperature (degrees Kelvin), ε_w (= 0.97) is the long-wave emissivity of the water surface (Edinger *et al.*, 1968), and ε_a (= 0.78 was employed depending on the estimation of Iziomon *et al.*, 2003) is the emissivity of the atmosphere. Note that T_s can be replaced by T assuming a well mixed condition and in the field, T was measured at the middle point of depth, near the river shore.

The evaporative heat transfer H_e can be estimated by the relation as follows:

$$H_e = -\rho L\, f(Wind)(e_s - e_a) \tag{22}$$

$$f(wind) = 0.0296 + 0.00637\ Wind_z \tag{23}$$

where e_a is the vapour pressure of the air, e_s is the saturated vapour pressure at the water surface temperature, L is the latent heat of water vaporization (= 2.45 MJ kg^{-1}), $f(Wind)$ is a wind function using wind velocity, $Wind_z$ is wind velocity at z height and z is vertical height (Gulliver & Stefan, 1986). Measured wind speed data was observed at 2 m height at the Hinheup station.

H_c is the convective heat transfer which may be expressed as follows:

$$H_c = 0.61 \frac{P}{1000} \rho L f(Wind)(T_a - T_s) \tag{24}$$

where T_a is the air temperature, P is atmospheric pressure and $f(Wind)$ is the same function of evaporative heat transfer.

The net heat exchange between the water and streambed S_b is expressed as:

$$S_b = -k \frac{\partial T_b}{\partial z} \tag{25}$$

where k is the thermal conductivity of the streambed materials, T_b is the streambed temperature and z is vertical distance into the streambed. Usually the riverbed heat conduction term S_b can be assumed to be negligible except under the extreme shallow flow condition (Honzo & Stefan, 1994). And in this study S_b was neglected because the flow depth was not so shallow and riverbed temperature data was not available.

The lateral advection term G can be calculated as follows:

$$G = \rho C p q_l (T_g - T_w) \tag{26}$$

where q_l is lateral inflow discharge per unit distance (m^3 s^{-1} m^{-1}) and T_g is lateral flow temperature. Usually lateral flow is supplied from groundwater. In this study, q_l was estimated by a rainfall–runoff model, which was analysed in a previous study (Yoshida *et al.*, 2004). The grand water temperature is almost the same as the annual average air temperature and does not change so much. Therefore, if the observed data was not available, T_g can be replaced as annual average air temperature (Bartholow, 1989).

Equation (17) was solved by the implicit Crank-Nikolson scheme using Δt = 3600 s and Δx = 1000 m.

RESULTS AND DISCUSSIONS

For model verification, the stream temperature was simulated with a 60% corridor vegetation density and a river discharge of 2 m^3 s^{-1}. Figure 7 shows the results at the observation point located 4 km downstream from Nam Song Dam, 17–22 March 2004. The daily maximum temperature values were almost in agreement with observed values; however, some minimum values were higher than the observed ones. This is because daily average wind speed data were used during the same day period; it should improve if hourly data were available. Figure 8 shows the comparison of input air temperature and simulated stream temperature 1–31 March 2004. Stream temperature change followed air temperature but its amplitude was smaller than air temperature.

Figure 9 shows the simulated temperature change along the Nam Song stream when the discharge is reduced from the pre-development value of 10 m^3 s^{-1} to the present value of 2 m^3 s^{-1} (the vegetation density is 60% in both cases). The vertical axis

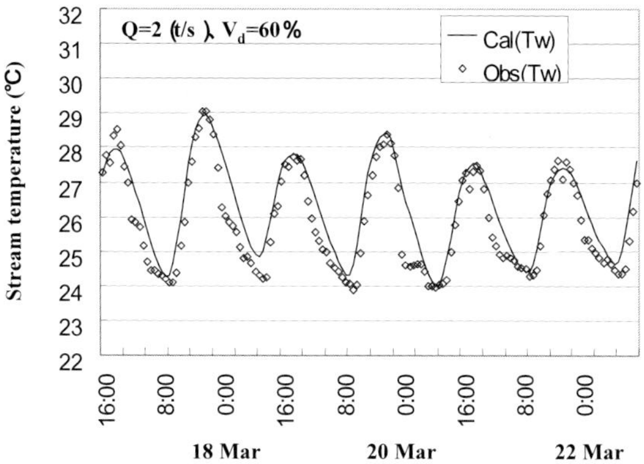

Fig. 7 Comparison of observed and calculated stream temperature (16–22 March 2004 at site A).

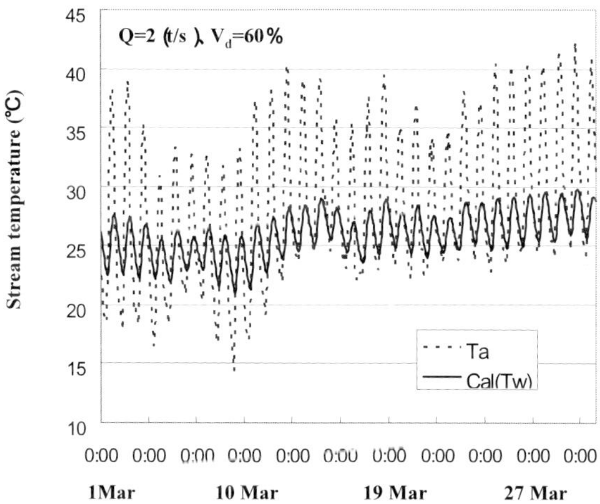

Fig. 8 Relation of observed air temperature (at Hinheup station) and calculated stream temperature (1–31 March 2004).

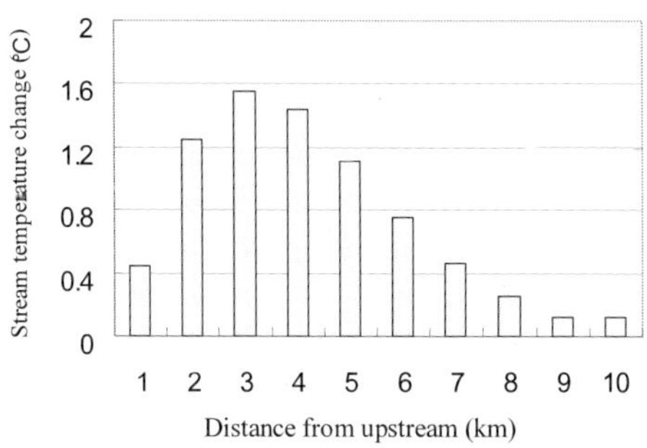

Fig. 9 Stream temperature changes along the stream between 10 $m^3 s^{-1}$ and 2 $m^3 s^{-1}$ discharge cases (from upstream boundary to 10 km downstream).

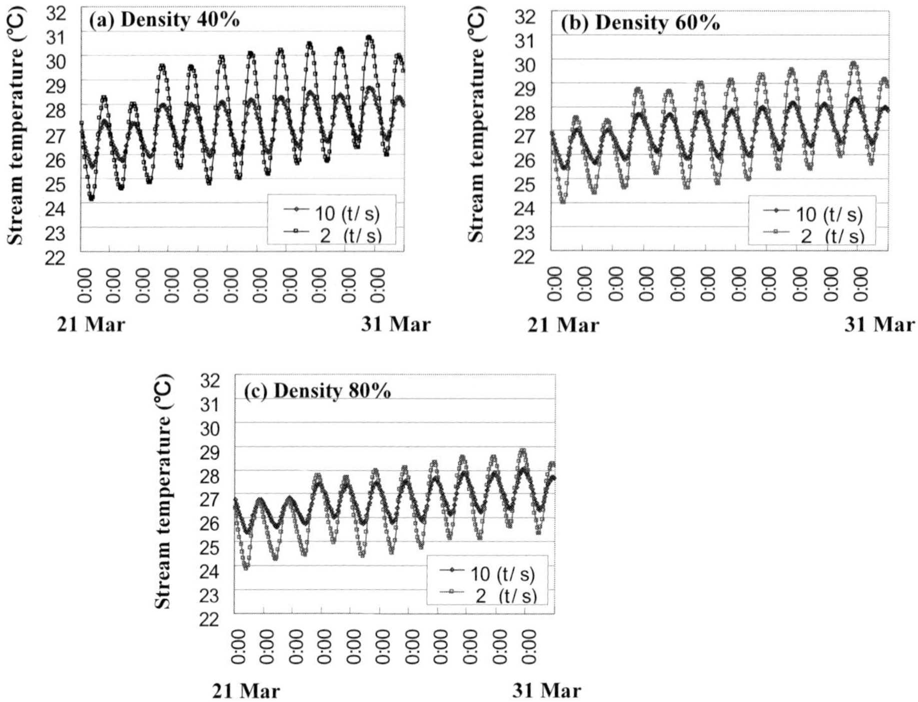

Fig. 10 Effect of density and discharge changes to the stream temperature (density: 40, 60 and 80%; discharge: 10.2 m^3 s^{-1}; at site A).

shows the change in monthly average temperature for March and the horizontal axis shows the distance from the upstream boundary. If we use one degree as the threshold of temperature change for the river ecological system, the zone between 2 and 5 km is subject to a temperature change that exceeds the threshold due to the reduction of discharge. A maximum temperature change of 1.6°C occurs at the 3-km point. The temperature below 6 km downstream is relatively unaffected because the discharge recovers downstream due to groundwater inputs.

Using the model, stream temperature was simulated with different corridor vegetation densities (40%, 60%, 80%) and discharges (10 m^3 s^{-1} for the pre-development and 2 m^3 s^{-1} for the present condition). Simulation results for 21–31 March 2004 are shown in Fig. 10(a)–(c). In the 10 m^3 s^{-1} discharge cases, the simulation results were within the one degree range in all vegetation density cases. However, in the 2 m^3 s^{-1} cases, stream temperatures increased by approximately two degrees as the vegetation density decreased. When vegetation density was high, stream temperature was not so sensitive to river discharge change. On the other hand, stream temperature was very sensitive to the discharge change when density was low. Therefore adequate management of river corridor vegetation is important to prevent the undesirable change of the river thermal environment.

CONCLUSION

In this study, firstly solar radiation was estimated considering daily and seasonal changes. Daily vegetation shade change was calculated from vegetation height, density and offset distance, and it was combined with a stream temperature analysis model. The model was applied to the Nam Song River to simulate the effect of corridor vegetation on stream temperature. Reducing the stream discharge from 10 $m^3 s^{-1}$ (pre-development condition) to 2 $m^3 s^{-1}$ (present condition) resulted in a change of up to 1.6° increase of stream temperature, strongly affecting the area between 2 km and 5 km from the upstream boundary. Simulation results using different vegetation density showed that stream temperature was higher and more sensitive to discharge change when the vegetation density was lower. Discharge change will occur with any water resource development. Therefore, the change in a river's thermal environment must be assessed before the development, and adequate management of river corridor is important to prevent the undesirable change of the river thermal environment.

Acknowledgements This research was partially supported by two Grant-in-Aids: Revolutionary Researches 2002 of the Ministry of Education, Science, Sports and Culture, and CREST of the Japan Science and Technology Corporation.

REFERENCE

Bartholow, J. M. (1989) Stream temperature investigation. Field and analytical methods. *Instream Flow Information Paper no. 13.* US Fish Wildlife Service Biological Report 89.

Bashar, A. S. & Stefan, H. G. (1993) Stream temperature dynamics (measurements and modeling). *Water Resour. Res.* **29**(7), 2299–2312.

Chen, Y. D., Carsel, R. F., McCutcheon, S. C. & Nutter, W. L. (1998) Stream temperature simulation of forested riparian areas: I. Watershed-scale model development. *J. Environ. Engng* **124**(4), 304–315.

Edinger, J. E., Duttweiler, D. W. & Geyer, J. C. (1968) The response of water temperature to meteorological conditions, *Water Resour. Res.* **4**(5), 1137–1143.

FAO (2000) *Crop Evapotranspiration.* FAO Irrigation and Drainage Paper no. 56. Food and Agriculture Organization, Rome, Italy.

Gulliver, J. S. & Stefan, H. G. (1986) Wind function for a sheltered stream. *J. Environ. Engng* **112**(2), 387–399.

Iziomon, M. G., Mayer, H. & Matzarakis, A. (2003) Downward atmospheric longwave irradiance under clear and cloudy skies: measurement and parameterization. *J. Atmospheric and Solar-Terrestrial Physics* **65**, 1107–1116.

LeBlanc, R. T., Brown, R. D. & FitzGibbon, J. E. (1979) Modeling the effects of land use change on the water temperature in unregulated urban stream. *J. Environ. Manage.* **49**(4), 445–469.

Roel, S. & Sean, W. (2001) Nam Song Diversion Project ADB TA5693 – Draft Impact Analysis Report and Action Plan, ADB.

Stanford, F. J. & Ward, J. V. (1988) The hyporheic habitat of river ecosystems. *Nature* **335**, 64–66.

Swift, L. W. & Messer, J. B. (1971) Forest cutting raise temperatures of small stream in the southern Appalachians. *J. Soil and Water Conservation* **26**, 111–115.

Yoshida, K., Tanji, H., Somura, H. & Toda, O. (2004) Evaluation of stream temperature change with reservoir management in Nam Ngum Basin. In: *Proc. ICID 2nd Asian Regional Conference* (Australia), session 3, stream 1, 1–8.

The hydrological and ecological effect of restoring the *Green Corridor* in the lower Tarim River, China

CONG ZHENTAO[1], NI GUANGHENG[1], LEI ZHIDONG[1] & MAHMUT BARATI[2]

1 *Department of Hydraulic and Hydropower Engineering, Tsinghua University, Beijing 100084, China*
congzht@tsinghua.edu.cn

2 *Tarim River Basin Administration Bureau, Korla 841000, China*

Abstract The downstream reach of the Tarim River (China), called the Green Corridor, has seriously deteriorated because of the drying-up of the Taitmar Lake and the downstream area since 1972. In order to restore the Green Corridor, water transfer has been undertaken on five occasions since May 2000 by diverting the Bosten Lake water into the Daxihaizi Reservoir and then making discharges into the downstream reach. Based on the monitoring data during the water transfer, the effects of stream flow on the rise in the groundwater table, soil moisture increase and vegetation recovery are studied. Water balance analysis indicates that with the planned amount of 0.35 billion m^3 water each year, the water requirements of the natural vegetation in a zone extending 2 km to each side of the stream channel can be met. The results are helpful in determining the proper timing and amount of water to be transferred.

Key words green corridor restoration; groundwater; inland river; soil moisture; vegetation cover; water balance; water transfer

INTRODUCTION

The Tarim River basin, with a catchment area of 1.02×10^6 km^2, is located in the Xinjiang Uygur Autonomous Region in northwest China. It is the largest inland river basin in China, and its mainstream and tributaries are shown in Fig. 1. The mainstream channel from Xiaojiake to the Taitmar Lake, 1321 km, is usually referred to as the Tarim River. It is a pure dissipative inland river with all the water resources supplied by its headstreams (Song Yudong, 2000).

In the past, water in the Tarim River could flow into the Taitmar Lake or even into the Lop Nur. The downstream reach of the Tarim River, from Qiala to the Taitmar Lake, with a length of 428 km, flowed between the Kuluketag Desert and the Taklimakan Desert. Because of the well developed vegetation cover along the river and its geographical and ecological significance, the downstream reach of the Tarim River is called the *Green Corridor*. However, with the reduction in water supply from headstreams and the low-efficiency exploitation of water and land resources in the upper and middle reaches, especially since the Daxihaizi Reservoir was built downstream for irrigation in 1972, no water now reaches downstream and the Taitmar Lake has dried up, resulting in the deterioration or even loss of the Green Corridor.

Fig. 1 Sketch map of the Tarim River basin (Z. D. Lei, 2001).

In order to rescue and restore the Green Corridor, water from the Bosten Lake is transferred to the Daxihaizi Reservoir and then discharged into the downstream reach of the Tarim River. Water transfer has been conducted on five occasions since May 2000, and the sixth is being undertaken. On the past five occasions, a total of 1.38×10^9 m^3 of water was transferred to the downstream reach based on the discharge record of the Daxihaizi Reservoir. Water flowed into the Taitmar Lake and formed a water surface area of around 200 km^2, which could be clearly identified on satellite images. At the same time, the groundwater table rose and the vegetation improved along both sides of the river.

A stream corridor is an ecosystem that usually consists of three major elements: stream channel, flood plain and transitional upland fringe (FISRWG, 1998). Given the importance of such corridors and the large number that have been destroyed, restoration research and practices have been conducted worldwide, such as the greenway planning in the USA (Fábos, 2004), the Savannah River Site in the USA (Bowers *et al.*, 2000), etc. However, the emphasis has mainly been concerned with humid areas, and seldom is research undertaken for arid areas where improvement of the hydrological conditions is the key to restoration.

Some research on the analysis and simulation of the water transfer (Cong *et al.*, 2002), dynamic change of groundwater levels (Xu *et al.*, 2003), plant communities (Liu *et al.*, 2002), and the influence of groundwater on vegetation (Chen *et al.*, 2003) has been carried out in association with the restoration of the downstream Tarim River. In this study, for the purpose of determining the proper timing and amount of water transfer, first the effects of stream flow on rising groundwater tables, soil moisture increase and vegetation recovery are analysed, and then water balance analysis for the periods of water transfer is conducted.

RESTORATION IMPLEMENTATION AND MONITORING

Tarim River and water transfer

The Tarim River divides at 4 km below the Daxihaizi Reservoir. The west branch is the old Tarim River which is 144 km long; the east branch is called the Qiwenkuoer River with a length of 205 km. The two branches join into one at Alagan, and then flow into the Taitmar Lake after travelling a distance of 153 km. The elevation difference between the Daxihaizi Reservoir and Taitma Lake is less than 50 m; the stream's longitudinal gradient decreases from upstream to downstream and the average gradient is as low as 0.12‰. The shape of the cross-sections is nearly a trapezium with the base width being 0.5 to 10 m and the side slopes 1:1 to 1:20.

In the alluvial plain of the downstream Tarim River, the main soil type is sand, including medium-fine sand, fine sand and silt sand. The groundwater is basically unconfined and the aquifer thickness varies from 50 to 100 m.

Water transfer has been undertaken on five occasions since May 2000, as summarized in Table 1.

Table 1 Summary of the five water transfers to the downstream Tarim River.

No	Period (Daxihaizi Discharge)	Last (day)	From Bosten Lake ($10^8 m^3$)	Daxihaizi discharge ($10^8 m^3$)	Flow distance (km)	Time to Taitmar Lake (day)
1st	2000.05.14－2000.07.12	60	2.04	0.99	106	
2nd	2000.11.03－2001.02.05	97	4.68	2.27	216	
3rd (1)	2001.04.01－2001.07.06	97	3.28	1.84	310	
3rd (2)	2001.09.12－2001.11.17	67	2.91	1.98	358	56
4th	2002.07.20－2002.11.10	113	4.38	3.31	358	77
5th	2003.03.03－2003.07.11	131	4.91	3.40	358+144*	20
Total			22.19	13.79		

*Note: The river length from Daxihaizi to Taitma Lake is about 358 km along the Qiwenkuoer River. In the case of double route water transfer, the river length of the old Tarim River, 144 km, should be added.

Monitoring

As shown in Fig. 2, to evaluate the hydrological and ecological effect, a total of 10 monitoring sites have been set up along the downstream reach of the Tarim River. At each site, surface water, groundwater, soil moisture and natural vegetation are observed at least once a month. Observations include water quality, vegetation growth, as well as discharge, river stage, groundwater table, soil moisture, etc.

HYDROLOGICAL AND ECOLOGICAL EFFECTS

Hydrological effect

Surface water Based on the monitoring data, discharges at each site are then calculated. Discharges at site "0", and "A" through "G" for the 2nd water transfer are

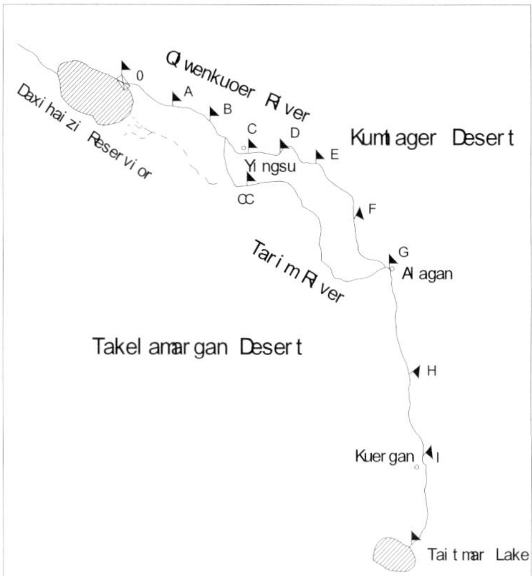

Note:

0-discharge gate of the Daxiahaizi Reservoir;

A-Akedong;

B-Yahepumahan;

C-Yingsu;

CC-oldYingsu;

D-Abudali;

E-Kaerdayi;

F-Tugemailai;

G-Alagan;

H-Yiganbujima;

I-Kuergan;

J-entrance of the Taitmar Lake.

Fig. 2 Monitoring sites along downstream of the Tarim River.

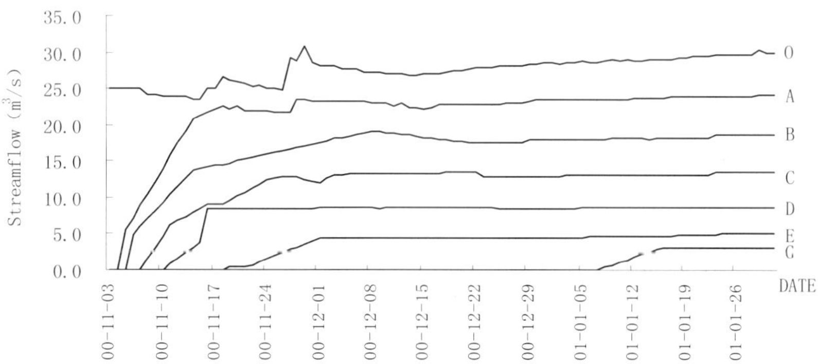

Fig. 3 Streamflow at each site during the second water transfer.

shown in Fig. 3; water loss in the flow direction along river can be seen clearly. Due to different initial conditions, water losses are also different for each water transfer.

River flow is mainly added by the discharge of the Daxihaizi Reservoir; this discharge and water loss along the river together determine how far the river flow can go. A sufficient amount of discharge is then necessary for the river flow to reach the Taitmar Lake.

Groundwater Since the groundwater level is lower, during water transfers river water recharges the groundwater. The recharge rate depends on river flow (stage), groundwater level, soil hydraulic properties, etc. The rise of groundwater level near the river will then induce groundwater flow away from both sides of the river channel, causing the groundwater table to rise. The groundwater table rises slowly where the distance to the river is great, from which the extent of the influence can be estimated.

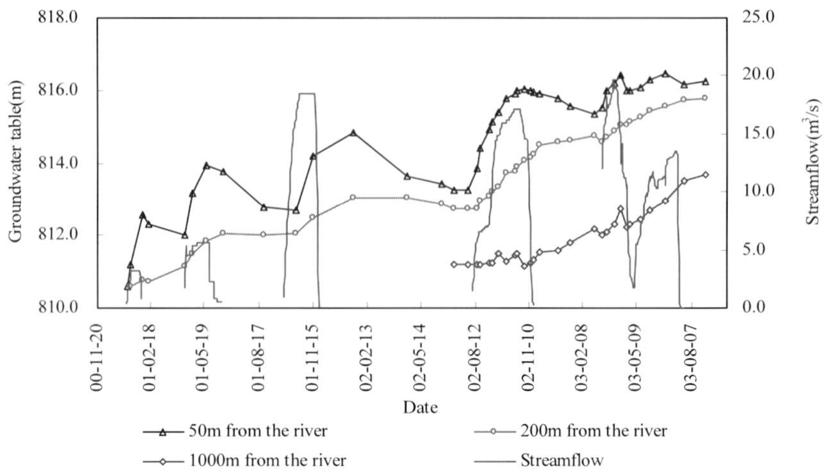

Fig. 4 Groundwater table change at the Alagan section.

For the five water transfer events, the extent of the influence is estimated as 1000 to 2000 m for both sides.

As illustrated in Fig. 4, in general the changes of groundwater table have a close relation to the river flow. At the monitoring well that is 50 m from the river, the groundwater table varies consistently with the river flow. When there is water flow in the river the groundwater table rises, and when river flow ceases the groundwater table starts to drop. However, there is a time lag between the changes of the groundwater table and river flow. With an increase of distance from the channel, the change of groundwater table becomes smooth. At the monitoring well 1000 m from the river, the groundwater table is found to rise up steadily.

Soil moisture Usually soil moisture is determined by soil type, evapotranspiration, groundwater level, etc. During water transfer, overflow and the rise of the groundwater table have important effects on soil moisture. As far as the overflow is concerned, because of the large uncertainty of its occurrence in time and space, it is rather difficult to observe its effects. Where there is no overflow, the soil moisture can be observed using the neutron probe method.

Figure 5 shows the soil moisture profile observed through time for the section at Alagan. From August to October 2002 during the fourth water transfer, soil moisture decreased continually at the site 450 m from the river, while when river discharge increased steadily to 17 $m^3 s^{-1}$ in February and March 2003, significant soil moisture increase occurred. Similar to groundwater, soil moisture is closely related with river flow, although the time lag of their changes is larger.

Ecological effect

Along with the changes in hydrological conditions, the ecology has been restored. The ecological improvement is judged mainly by the extending of the community or the

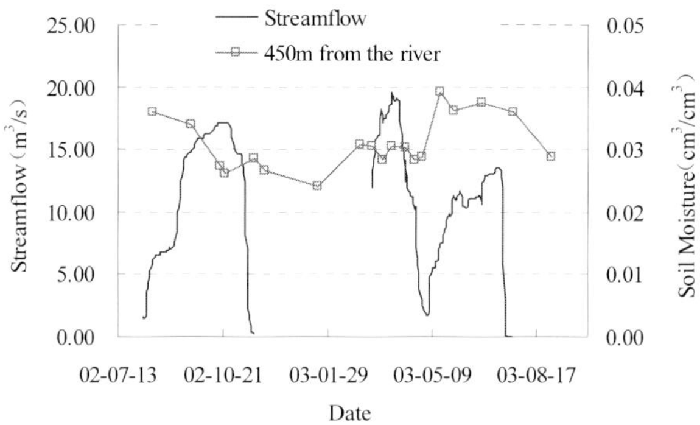

Fig. 5 Soil moisture and river flow at the Alagan section.

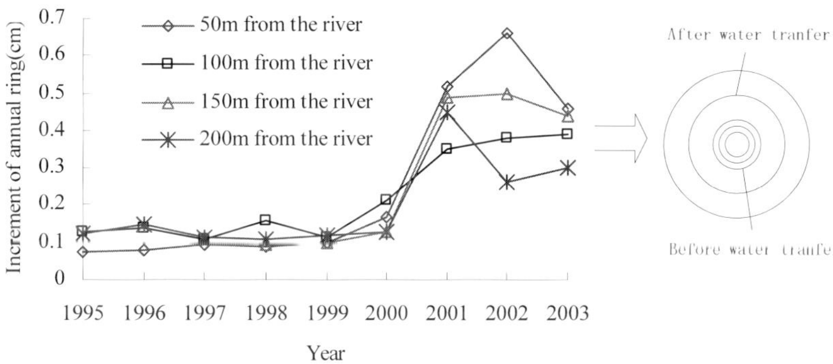

Fig. 6 Change in the annual ring growth of *Populus euphratica* at Yingsu.

Table 2 Land cover distribution before and after water transfer (km^2).

Type of land cover	August 1999	August 2002
Farmland	90.7	99.1
Natural vegetation	1963.6	2143.5
Water surface	6.8	131.4
Sand land	5598.3	5261.3
River	0.0	24.1
Total	7659.3	7659.3

growth of individual plants. The annual growth rings of *Populus euphratica* are selected as representative of the growth of individual trees (Water Resources Bureau of Xinjiang, 2004). The increment of the annual ring is smaller before water transfer and becomes larger after water transfer, as illustrated in Fig. 6.

By satellite image analysis of the same season but for different years, the change of the vegetation community can be estimated and the result is shown in Table 2 (Tarim River Basin Administration Bureau, 2001). After water transfer, the area of both water surface and natural vegetation increased significantly, while a large decrease was found in the sand land area.

WATER BALANCE ANALYSIS

The final aim of the water transfer is to restore the vegetation in the Green Corridor, which has a close relationship with the hydrological conditions. In the lower reach of the Tarim River, precipitation is rare and the potential evapotranspiration is strong. At Tieganlike, the mean annual precipitation is 36.3 mm while the mean annual evaporation reaches 2517.6 mm. The annual average temperature of the area is 11.6°C and the temperature difference in a day is large. The sunshine duration is as high as 2890 to 3090 hours in a year. Therefore the hydrological conditions in terms of groundwater level, soil moisture, etc. have a great influence on the vegetation. Based on the monitoring data during the water transfer, a water balance analysis was carried out for different reaches; the items considered include groundwater recharge, soil moisture increase and evapotranspiration, etc.

From May 2000 to July 2003, the total water discharge from the Daxihaizi Reservoir was about 1.379×10^9 m^3. It is important to know where the water goes, which is helpful to evaluate the effects of water transfer and to make a plan for water transfer. The water balance equation can be written as:

$$W_t + P = W_g + W_s + ET + W_{lake}$$

W_t is total water transfer; P is precipitation; W_g is groundwater recharge; W_s is water supply to soil; E_t is evapotranspiration; W_{lake} is water flowing into lake.

The groundwater recharge from surface water is reflected by the groundwater table rise. Therefore, based on the observed groundwater table of each section, the specific yield, and the length of river reach, the groundwater recharge can be calculated. From May 2000 to July 2003, the area-averaged groundwater table rise was 2.5 m along the Qiwenkuoer River, and 1.4 m or greater along the old Tarim River. The groundwater change extended about 1500 m laterally on both sides of the river upstream of Alagan, and 1000 m downstream of Alagan. By using the above method, groundwater recharges of 0.561×10^9 m^3 along Qiwenkuoer River and 0.104×10^9 m^3 along the Old Tarim River are estimated. Of the total of 1.379×10^9 m^3 water from the Daxihaizi, the total amount of groundwater recharge reaches 0.665×10^9 m^3.

The water discharged into the Taitmar Lake was about 0.045×10^9 m^3. Since the precipitation is very small and may be ignored, the evapotranspiration is then estimated as 0.660×10^9 m^3 by using the water balance equation. Therefore, of the total water outflow from the Daxihaizi Reservoir, about 50% was used to supply the groundwater and soil water and the other 50% was used as evaporation and transpiration.

CONCLUSION AND DISCUSSION

With more and more water flowing into the lower Tarim River, the groundwater table will continue to rise, which may then bring some impacts to the Green Corridor. The area of restored vegetation will increase; the evaporation and transpiration will increase; at the same time more water will be needed to raise or maintain the groundwater table. In the case that the transferred water is limited, the groundwater table will reach a stable level and the water will be consumed through evaporation and transpiration.

Based on satellite data, along the lower Tarim River in a zone 2 km from the river, the area of natural vegetation is about 870 km^2. If the actual evapotranspiration of the natural vegetation is assumed to be 400 mm $year^{-1}$, the water required will be 0.35×10^9 m^3 per year, which is the same amount as defined in the government approved water transfer plan, i.e. discharge from the Daxihaizi Reservoir is controlled at 0.35×10^9 m^3. Therefore, it is concluded that the possible range of the vegetation restoration will be within 2 km of the river on both sides.

REFERENCES

Bowers, C. F., Hanlin, H. G., Guynn Jr, D. C., McLendon, J. P. & Davis, J. R. (2000) Herpetofaunal and vegetational characterization of a thermally-impacted stream at the beginning of restoration. *Ecological Engineering* **15**, Suppl. 1, S101–S114.

Chen Yangning, Li Weihong, Xu Hailiang, *et al.* (2003) The influence of groundwater on vegetation in the lower reaches of Tarim River, China. *Acta Geogrophica Sinica* **58**(4), 542–549.

Cong Zhentao, Zhou Haiying, Lei Zhidong, Wu Liuyi & Omerjan (2003) Analyses and simulations of the water transfer from the lower Tarim River. *Adv. Water Sci.* **14**(3), 276–279.

Fábos, J. G. (2004) Greenway planning in the United States: its origins and recent case studies. *Landscape and Urban Planning* **68**(2–3), 321–342.

FISRWG (Federal Interagency Stream Restoration Working Group) (1998) *Stream Corridor Restoration: Principles, Processes and Practices.* National Technical Information Service, US Department of Commerce, Springfield, Virginia, USA.

Lei, Z. D., Zhen, B. L., Shang, S. H., Yang, S. X., Cong, Z. T., Zhang, F. W., Mao, X. H. & Zhou, H. Y. (2001) Formation and utilization of water resources of Tarim River. *Science in China* **44**(6), 615–624.

Liu Jiazheng & Chen Yaning (2002) Analysis on converse succession of plant communities at the lower reaches of Tarim River. *Arid Land Geography* **25**(3), 232–236.

Song Yudong, Fan Zili, Lei Zhidong & Zhang Fawang (2000) *Research on Water Resources and Ecology of Tarim River, China.* Xinjiang People's Press, China.

Tarim River Basin Administration Bureau (2001) Analysis on the ecological and environment effect of the water transfer in Green Corridor of Tarim Downstream.

Xu Hailiang, Song Yudong & Chen Yaning (2003) Dynamic change of groundwater after ecological water transport at the lower reaches of Tarim River. *China Environ. Sci.* **23**(3), 327–331.

Water Resources Bureau of Xinjiang (2004) Research on monitoring and evaluation of the water transfer and ecological improvement in Tarim Downstream.

Dynamics and Biogeochemistry of River Corridors and Wetlands (Proceedings of symposium S4 held during the Seventh IAHS Scientific Assembly at Foz do Iguaçu, Brazil, April 2005). IAHS Publ. 294, 2005.

Flooding-cycle analysis in wetlands with negligible overland drainage

HUGO A. LOÁICIGA[1] & ALLISON HUANG[2]

1 *Department of Geography, University of California, Santa Barbara, California 93106, USA*
hugo@geog.ucsb.edu

2 *Bren School of Environmental Science and Management, University of California, Santa Barbara, California 93106, USA*

Abstract This paper presents a methodology to calculate the ponding and infiltration hydrographs in rainfed wetlands with negligible overland drainage. The flooding cycle is made up of at most three phases. In the first, the rainfall rate equals the infiltration rate and there is no ponding. The second phase begins with the initiation of ponding and ends with the end of rainfall or of ponding, whichever comes first. The third, and last, phase starts with the end of rainfall—provided that there is ponded water still extant—and ends with the extinction of ponding. The ponding and infiltration hydrographs are calculated by solving ordinary differential equations of water balance and soil-water flow. Computational examples demonstrate the methodology developed in this work using soils with very different hydraulic properties.

Key words differential equations; infiltration; rainfall; soil moisture; wetlands

INTRODUCTION

On the nature of wetlands

Wetlands are generally defined as areas covered intermittently or permanently by shallow waters, or underlain by saturated, hydric (poorly-drained) soils during prolonged periods. Wetlands include a variety of landscapes: bayous, bogs, fens, mangroves, marshes, potholes, river overbanks, sloughs, swamps, vernal pools, and wet meadows (Hammer, 1997; Kent, 2001). They may contain fresh or brackish waters and are found in any landscape with a local natural source of water and suitable topography, climate, and soils or rocky substrate. If the water source of a wetland is not natural or if the land surface has been modified, the result is a created wetland (Hammer, 1997). When the land surface is reshaped to an original condition or if a natural water source is re-established, the result is a restored wetland (Rosgen, 1996). Permanent streams, reservoirs, and deep lakes are not considered wetlands and are generally devoid of rooted plants. Wetlands play important ecological and hydrological functions in a wide variety of landscapes, and justification for their preservation and restoration has been established on economic and ecological grounds (Loáiciga, 2001). Brinson (1995) proposed a well-known hydrogeomorphic classification for wetlands.

Research objective

The following sections contain a novel methodology to quantify the temporal evolution of infiltration and ponding in rain-fed wetlands with negligible overland drainage. The methodology focuses on mechanistic quantitative analysis of wetland ponding caused by design storms (of known duration and temporal intensity distribution, and specified return interval). The methodology and related research findings are useful in determining landscape changes to improve the water-retention properties in seasonal wetlands and in wetland restoration in general. It is assumed in the following analysis that the water table is deep and does not interfere with infiltration. Yet, the calculated infiltration can be used to assess localized potential for deep groundwater recharge.

THE FLOODING CYCLE IN RAIN-FED WETLANDS

The flooding cycle of wetlands considered in this work is triggered by design storms of specified: (a) duration (typically 24 h or shorter); (b) temporal intensity distribution; and (c) return interval (two years or longer). These storms deliver water with sufficient intensity over short times so as to have the potential to pond water on terrain with negligible overland drainage. The wetlands considered in this work are found in regions with Mediterranean-type climate, typified by well-defined wet and dry seasons, the former featuring a limited number of storms during the cool winter months (December to March) that account for all or most of the yearly rainfall, and the latter having potential evapotranspiration well in excess of seasonal rainfall during most of the warm, non-winter, months. Specifically, vernal pools in coastal mesas and seasonal meadows in inland valleys of central and southern California are prototypes of the wetlands considered in this work. The flooding cycle in this type of wetland has a well-defined sequence of phases for a design storm of duration D that starts at time $t = 0$, followed by rainfall that increases monotonically until time t_{peak}, and then declines until vanishing at time D. In the first phase of a wetland's flooding cycle all the rainfall infiltrates into the wetland's soil. The second phase starts with the initiation of ponding at time t_p. The rainfall rate exceeds the infiltration rate in this phase, and ponding increases for as long as the rainfall rate exceeds the infiltration rate. At some time (t_{max}) the rainfall rate becomes equal to the infiltration rate. The ponding depth is maximal at time t_{max}. Ponding recedes after time t_{max}, and two situations can arise during its recession. In the first—typical of highly permeable soils—ponding vanishes before the end of rainfall, thus ending the wetland's flooding cycle. In the second, there is ponded water at the end of rainfall (at time = D), giving rise to the third phase. The third phase begins at time D, when evaporation of ponded water also starts. Ponding recession continues by the combined action of infiltration and evaporation until it is completely depleted, at time t_f, the extinction time of ponding. The processes of infiltration and ponding just described can be quantified with specialized water-balance equations developed in the next section.

Several additional observations are pertinent concerning the flooding and drying cycles of wetlands considered in this work. The first is that transpiration of infiltrated water by vegetation is negligible during the rainy, cool, season in our study region. This is so because vegetation—characterized in this case by seasonal grasses and weeds—is dead or dormant. Infiltration and evaporation are therefore the two main mechanisms responsible for the depletion of ponded water in the wetlands. Second, the maximum ponded depth Y_{max}, which occurs at time t_{max}, and the duration of ponding ($D_p = t_f - t_p$) are variables of particular interest in this work. Y_{max} defines—for a given design storm—the minimum height of lateral containment to be provided on a wetland's perimeter so as to maximize water retention in the restoration of degraded wetlands. It is also useful in calculating flooding area, which is determined from Y_{max} and the elevation of the land surrounding the wetland's floor. In the type of wetlands considered in this work the flooding area varies from a few hundred square metres to a few hectares (1 ha = 10^4 m^2). The duration of ponding (D_p) is valuable in the assessment of ecological and recreational function of seasonal wetlands, and in determining the need for artificial water inputs to lengthen the duration of ponding if so desired.

WETLAND WATER BALANCE WITH EMPIRICAL INFILTRATION

There are several empirical models of infiltration capacity, see Rawls *et al.* (1996) for a review. The use of these empirical models in wetland-ponding analysis can be illustrated with the well-known model due to Horton (1940), in which the infiltration capacity, $f_H(t)$, in units of length per time, is given by a three-parameter, exponential, function:

$$f_H(t) = f_c + (f_0 - f_c)e^{-kt} \quad t \geq 0 \tag{1}$$

where f_c is the asymptotic infiltration capacity, f_0 denotes the initial infiltration capacity (at time $t = 0$) , and k is a time-decay constant (dimensions of 1/time). The Horton parameters (f_0, f_c, k) must be estimated from infiltration measurements, usually using infiltrometers. The (actual) infiltration rate, f, equals the infiltration capacity if the rainfall rate w equals or exceeds $f_H(t)$. The theoretical cumulative infiltration, $F^*(t)$ in the period [0, t] is the integral of equation (1) evaluated between 0 and t:

$$F^*(t) = t \cdot f_c - \frac{f_H(t) - f_0}{k} \tag{2}$$

The depth of ponding at the end of the period [0, t], $Y(t)$, during which $w \geq f_H(t)$ is:

$$Y(t) = \int_0^t [w(u) - f_H(u)]\mathrm{d}u = W(t) - F^*(t) \tag{3}$$

in which $W(t)$ denotes cumulative rainfall, and the theoretical cumulative infiltration $F^*(t)$ is given by equation (2).

In actuality, storms in general, and design storms in particular, have low intensity during their early stages. The infiltration capacity exceeds the rainfall rate in this period, and the actual infiltration rate equals the rainfall rate, leaving no rainfall excess for ponding during this early phase. Infiltration at a rate less than the infiltration capacity complicates the implementation of the infiltration capacity models of the type exemplified by equations (1) and (2). An iterative procedure is required to determine the time at which ponding starts (t_p) using empirical infiltration models of the Horton type. The time to ponding (t_p) is that time that meets two conditions: (a) the actual cumulative infiltration equals the cumulative rainfall at t_p, and (b) the initial infiltration capacity (f_p, in equation (4), below) equals the rainfall rate at t_p. The rainfall rate exceeds the infiltration capacity after t_p. The infiltration capacity during this (ponding) phase is given by:

$$f_{pH}(t) = f_c + (f_p - f_c)e^{-k(t-t_p)} \qquad t \geq t_p \tag{4}$$

The ponding phase using empirical infiltration

After t_p ponding increases monotonically until time $t_{\max}$, when the maximum ponding depth, $Y_{\max}$, is:

$$Y_{\max} = \int_{t_p}^{t_{\max}} [w(u) - f_{pH}(u)]\mathrm{d}u = W(t_p, t_{\max}) - F^*_{pH}(t_p, t_{\max}) \tag{5}$$

In the period $t > t_{\max}$ the rainfall rate is less than the infiltration capacity. Yet, due to the presence of ponded water, the infiltration rate equals the infiltration capacity $f_{pH}(t)$. Ponding subsides in the period $t > t_{\max}$ and may either vanish altogether prior to the end of rainfall—which occurs at time D—or continue to recede beyond D. In the latter instance, evaporation (e_v, a known input function) commences at time D and, in conjunction with infiltration, contributes to ponding recession. The ponding depth at the end of rainfall, Y_D, equals:

$$Y_D = \int_{t_p}^{D} [w(u) - f_{pH}(u)]\mathrm{d}u = W(t_p, D) - F^*_{pH}(t_p, D) \tag{6}$$

The time of ponding extinction (t_f) is obtained by solving the following integral equation:

$$W(t_p, D) - \int_{D}^{t_f} e_v(u)\,\mathrm{d}u - F^*_{pH}(t_p, t_f) = 0 \tag{7}$$

in which the unknown is the time of ponding t_f.

In practice, there are difficulties implementing empirical infiltration models of the Horton type. The estimation of the Horton parameters, in particular, is far from trivial, even when using simple devices such as the double ring infiltrometer. One of the main hurdles arises from the difference between the pattern of rainfall falling on soils and the instantaneous ponding of water created inside an infiltrometer at the beginning of, and maintained throughout, an infiltration test. The latter measurement method is likely to overestimate the infiltration capacity. Figure 1 shows measured infiltration capacities in Sorrento clay in a seasonal meadow in Ojai, California, USA. Measurements were made using a double-ring infiltrometer in moist conditions following the first autumn storms in November 2003, and during dry conditions in the summer (August) of 2004. The two infiltration curves differ in the early and late stages of water application, a phenomenon attributed to changes in the structure of this expansive clay caused by a varying water content. Changes in clay structure, in turn, affect its

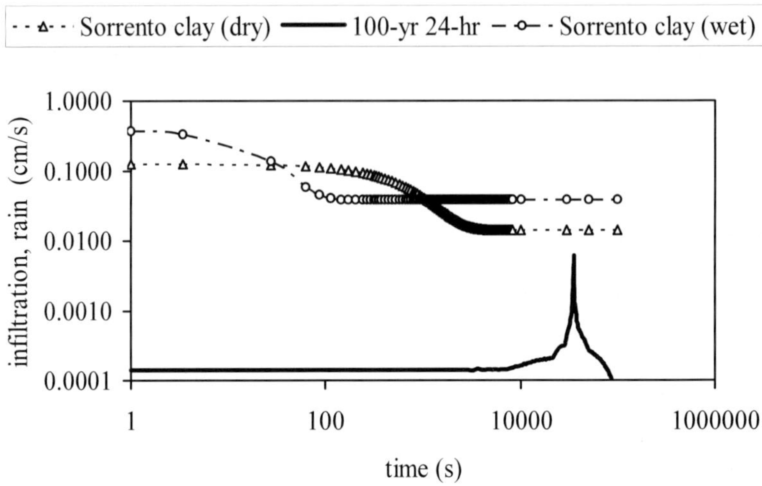

Fig. 1 Hortonian infiltration for Sorrento clay under dry and moist conditions, and the 100-year, 24-h, hyetograph.

hydraulic properties, and, thus, its infiltration capacity. Also shown in Fig. 1 is the hyetograph for the 100-year, 24-h storm in Ojai, California. Evidently, the infiltration capacities exceed the rainfall intensities, implying that ponding is not possible under either type of antecedent condition prevailing during measurements. Another important drawback of empirical infiltration models is that they do not incorporate mass-conservation and flow equations in their mathematical description of infiltration. The next section introduces a physically-based infiltration model in the analysis of the flooding cycle of wetlands.

FLOODING CYCLE USING PHYSICALLY-BASED INFILTRATION

The Green-and-Ampt method (Green & Ampt, 1911) relies on an analysis of rainfall, infiltration, and evaporation that incorporates conservation of mass and Darcy's law in the calculation of actual infiltration rate (f) in a vertical column of a homogeneous soil with an initial moisture deficit (equal to the soil porosity minus the initial volumetric water content, $n - v_0$, a constant soil-water tension at the downward-advancing saturation front (h_f, where $h_f > 0$), a soil property that can be estimated from a soil's textural properties, Rawls *et al.*, 1996), and a saturated hydraulic conductivity (K_0). A saturation front at a depth z_f (measured vertically from the ground surface, $z_f < 0$) in a soil column of unit cross-sectional area implies a cumulative infiltration $F(t) = |z_f(t)|(n - v_0))$. During the pre-ponding phase the rainfall rate (w) equals the infiltration rate (f), so that cumulative infiltration equals cumulative rainfall, $F(t) = W(t)$. The magnitude of the depth of the saturation front at the initiation of ponding, $|z_f(t_p)|$, is unknown. To obtain it one writes Darcy's law between the ground surface (i.e. at $z = 0$, where the hydraulic head is zero) and $z_f(t_p)$ (where the hydraulic head equals $z_f(t_p) - h_f$) and sets the magnitude of the Darcian flux equal to the rainfall rate at time t_p (i.e. $w(t_p)$). Solving for $|z_f(t_p)|$ produces $|z_f(t_p)| = K_0 h_f / (w(t_p) - K_0)$, with $w(t_p) > K_0$ required, otherwise ponding does not occur. Using the last expression for $|z_f(t_p)|$ in the conservation equation $F(t) = |z_f(t)|(n - v_0)$ produces the following integral equation for the time to ponding t_p:

$$W(t_p) = K_0 h_f \cdot (n - v_0)/[w(t_p) - K_0] \tag{8}$$

in which the rainfall rate $w(t_p)$ is known. Following the onset of ponding and until the end of rainfall, the cumulative rainfall is partitioned into cumulative infiltration and cumulative ponding, $W(t) = F(t) + Y(t)$, for $t_p \leq t \leq D$. To obtain the cumulative infiltration $F(t)$, or its derivative, the actual infiltration rate, $f(t) = F'(t)$, Darcy's law is written between the ground surface (i.e. at $z = 0$, where the hydraulic head equals $Y(t) = W(t) - F(t)$) and $z_f(t)$ (where the hydraulic head equals $z_f(t) - h_f$), and the magnitude of the Darcian flux is set equal to the actual infiltration rate. Furthermore, $F(t) = |z_f(t)|(n - v_0)$, is established above. This equation is combined with Darcy's law to produce the following equation for the cumulative infiltration $F(t)$ in the interval $t_p \leq t \leq D$:

$$F\ F' - vF - c \cdot (W(t) + h_f) = 0 \tag{9}$$

whose initial condition is $F(t_p) = w(t_p)$, in which the cumulative rainfall $W(t)$ is a known function, and $v = K_0 \cdot (1 - (n - v_0))$, $c = K_0 \cdot (n - v_0)$. Equation (9) governs the evolution of infiltration from the onset of ponding until the end of rain or the cessation of ponding, whichever comes first. Its solution is efficiently arrived at numerically using the Runge-Kutta method (see, for example, Carnahan *et al.*, 1969). Ponding is reduced by infiltration and evaporation after rainfall. Evaporation (e_v, a known input function) starts at $t = D$, i.e. at the end of rainfall. The water balance equation in the period $t \geq D$ is:

$$W(D) = Y(t) + F(t) + \int_D^t e_v(u)\,du = Y(t) + F(t) + E(D,t) \tag{10}$$

To obtain the cumulative infiltration $F(t)$, Darcy's law is written between the ground surface (i.e. at $z = 0$, where the hydraulic head equals $Y(t) = W(D) - F(t) - E(D, t)$) and $z_f(t)$ (where the hydraulic head equals $z_f(t) - h_f$), and the magnitude of the Darcian flux is set equal to the actual infiltration rate. Furthermore, $F(t) = |z_f(t)|(n - v_0)$. This equation is combined with Darcy's law to produce the following equation for the cumulative infiltration $F = F(t)$ when $t \geq D$:

$$F\ F' - vF - c \cdot (W(D) + h_f - E(D,t)) = 0 \tag{11}$$

whose initial condition is $F(D)$ (from the solution of equation (9)) and $v = K_0 \cdot (1 - (n - v_0))$, $c = K_0 \cdot (n - v_0)$. The evaporation in equation (11) is an external input.

Figure 2 shows the cumulative rainfall (100-year, 24-h storm) in Ojai, California, and the resulting cumulative infiltration (F), and ponding (Y) in a wetland underlain by silt-clay soil ($K_0 = 1.03 \times 10^{-4}$ cm s^{-1}). The calculated maximum ponding depth and the duration of ponding were 14.09 cm and 153 546 s, respectively. Figure 3 shows, for the sake of comparison, analogous information to that in Fig. 2 for the same location but in a wetland underlain by silt loam ($K_0 = 7.20 \times 10^{-4}$ cm s^{-1}). The relatively large saturated hydraulic conductivity of silt loam delays the emergence of ponding in this soil when compared with silt clay. In addition, ponding is short lived in silt loam, and, in fact, ponding ends before the termination of rainfall. The maximum calculated ponding depth and ponding duration in silt loam were 3.31 cm and 7002 s, respectively. The calculations used to construct the graphs in Figs 2 and 3 were carried with a Runge-Kutta solver for ordinary differential equations coded in the commercial software MATLAB.

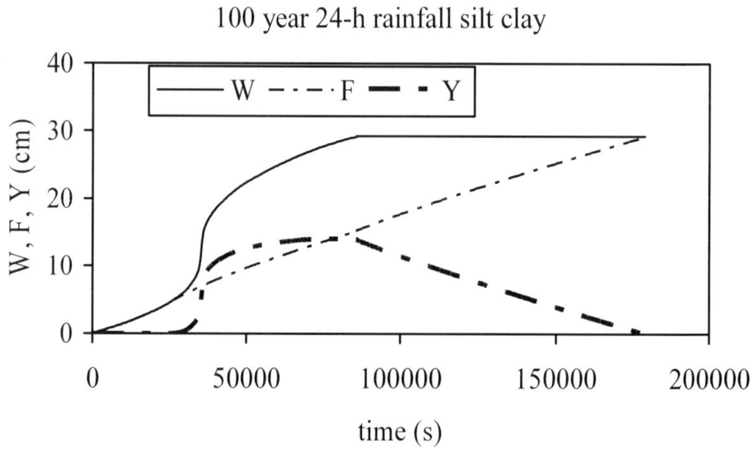

Fig. 2 Cumulative rainfall (W), infiltration (F), and ponding (Y) caused by the 100-year, 24-h, rainfall in a seasonal meadow underlain by silt clay. $Y_{max} = 14.09$ cm. Ponding duration $D_p = 153\ 546$ s.

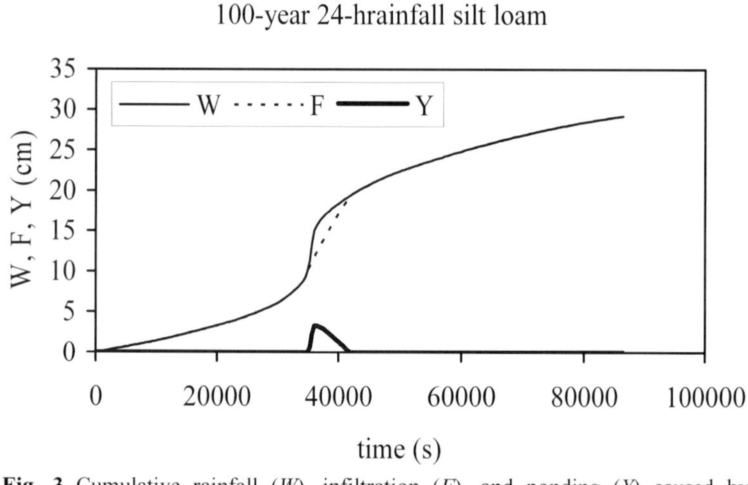

Fig. 3 Cumulative rainfall (W), infiltration (F), and ponding (Y) caused by the 100-year, 24-h, rainfall in a seasonal meadow underlain by silt loam. $Y_{max} = 3.31$ cm. Ponding duration $D_p = 7002$ s.

REFERENCES

Brinson, M. M. (1995) The HGM approach explained. *Natural Wetlands Newsletter,* November/December, 7.

Carnahan, B., Luther, H. A. & Wilkes, J. O. (1969) *Applied Numerical Methods*. John Wiley & Sons, New York, USA.

Green, W. H. & Ampt, G. A. (1911) Studies on soil physics. 1: The flow of air and water through soils. *J. Agric. Sci.* **4**, 1–24.

Hammer, D. A. (1997) *Creating Freshwater Wetlands*, 2nd edn. Lewis Publishers, Boca Raton, Florida, USA.

Horton, R. E. (1940) An approach toward a physical interpretation of physical infiltration capacity. *Soil. Sci. Soc. Am. Proc.* **5**, 399–417.

Kent, D. M. (2001) *Applied Wetlands Science and Technology*. 2nd edn. Lewis Publishers, Boca Raton, Florida, USA.

Loáiciga, H. A. (2001) Flood damages under changing floodplain conditions: a forensic-hydrology case study. *J. Am. Water Resour. Assoc.* **37**(2), 467–478.

Rawls, W. J., Goldman, D., Van Mullen, J. A. & Ward, T. J. (1996) Infiltration. In: *Hydrology Handbook*, 75–124. Am. Soc. Civil Engineers, New York, USA.

Rosgen, D. (1996) *Applied River Morphology*, 2nd edn. Wildland Hydrology, Pagosa Springs, Colorado, USA.

Dynamics and Biogeochemistry of River Corridors and Wetlands (Proceedings of symposium S4 held during the Seventh IAHS Scientific Assembly at Foz do Iguaçu, Brazil, April 2005). IAHS Publ. 294, 2005.

Potential ecohydrological controls on peat degradation and vegetation pattern change in a kettle-hole bog

J. M. WADDINGTON, M. J. T. HEYWOOD, B. D. CROSBIE & E. C. DOWSETT

School of Geography and Geology, McMaster University, Hamilton, Ontario L8S 4K1, Canada
wadding@mcmaster.ca

Abstract Over the last 200 years land-use change in southern Ontario, Canada, has reduced wetland area and altered peatland vegetation. More recent threats to wetlands include enhanced nutrient loading from agricultural areas, which may have a profound impact on bog vegetation and carbon dynamics—potentially leading to peat degradation. We examined vegetation pattern change (1964–1999) and contemporary carbon dioxide exchange in a kettle-hole bog in an agricultural landscape undergoing a shift to crops with nonpoint source nutrient loading. Many of the zones of bog vegetation change (loss in *Sphagnum* sp. and *Chamaedaphne calyculata* and increase in *Carex* spp. and *Thelypteris palustris*) occurred near the lagg in zones that also had high NO_3^- concentrations (10–20 mg l^{-1}). The loss of typical bog species along the margin of the peatland (adjacent to the lagg) was accompanied by an expansion of the lagg from 11 to 17% of the overall peatland area. Carbon dioxide exchange results suggest that some sections of the bog are degrading (large loss of CO_2 to the atmosphere). Although it has been suggested that watershed deforestation by European settlement led to the expansion of southern Ontario kettle-hole bogs, the patterns of vegetation change in this study suggest that this bog is reverting to a poor fen.

Key words agriculture; carbon dioxide exchange; GIS; hydrology; nitrate; vegetation; wetland

INTRODUCTION

Kettle-hole peatland development is a function of both autogenic (natural and internal) and allogenic (external forces, such as human and climate) processes (Bunting *et al.*, 1998). Warner *et al.* (1989) suggest that the formation of many kettle-hole bogs in the Great Lakes drainage basin (North America) may have been caused by human activity as the period of rapid horizontal growth of *Sphagnum* mats corresponds with the period of rapid deforestation during European settlement (~200 years BP) and the associated changes in watershed hydrological conditions (Warner *et al.*, 1989). Recently, due to the application of nitrogen fertilizer, groundwater in many southern Ontario agricultural landscapes has become contaminated by nitrate (NO_3^-). While wetlands have a high potential to reduce natural and anthropogenic inputs of NO_3^- (Hill, 1991), excess NO_3^- entering a bog through agricultural runoff has the potential to change the biogeochemistry of the bog if the bog is not isolated hydrologically. Changes in peatland nutrient conditions are particularly important because peatland vegetation is strongly influenced by nutrient dynamics and carbon dynamics can also be altered.

Despite mounting evidence of the potential impact of increased NO_3^- loading on vegetation dynamics in southern Ontario kettle-hole bogs, this issue has not been studied directly. The objectives of this study, therefore, were to: (a) determine the change in vegetation patterns, 1964–1999, in a kettle-hole bog in southern Ontario, (b) determine whether the peatland is degrading (net source of atmospheric CO_2), and to (c) examine the potential ecohydrological controls (e.g. moisture, nutrient cycling, flow pathway) on vegetation pattern change and CO_2 exchange.

MATERIALS AND METHODS

Study site

The study was conducted at the Copetown Bog, an 11 ha provincially significant kettle-hole bog, near Hamilton, Ontario, Canada (43°17′N, 79°59′W). Agricultural land in the bog's watershed decreased by over 26% between 1960 and 1990; however, there was a shift from the production of hay and grass for livestock grazing to cash crops such as soybean and corn. Forest area increased by over 14% and urban areas increased by almost 7% of the watershed area. The bog lagg increased from 11% to 17% of the total bog area.

Copetown Bog has a raised ombrotrophic centre surrounded by a shallow (<50 cm) minerotrophic lagg (McHattie, 1992; Fig. 1). The bog is surrounded by agricultural land and a recently (around 1990) dredged small pond to the south. The surface of the bog has a typical peatland hummock (20–50 cm) and hollow (<15 cm deep) micro-topography. The species dominating these areas include *Chamaedaphne calyculata*, *Carex* spp. and *Sphagnum* mosses. The lagg vegetation is far more diverse, including *Juncus* sp., *Thelyptris palustrus*, *Maianthemum canadense*, *Calla palustris*, *Scirpus* sp., *Typha latifolia*, *Chamaedaphne calyculata* and *Ledum groenlandicum*.

Bog vegetation mapping

The presence and absence of vegetation was initially mapped in 1964 (Radforth, 1964) but not all plants were identified to the species level. Therefore, to facilitate meaningful comparison with the previous study we identified vegetation to the same taxonomic level. Vegetation mapping was carried out weekly during the months of June to September 1999. Identifications were verified using Newmaster *et al.* (1997). Vegetation coverage maps, illustrating the zones of appearance and disappearance of vegetation genera and species between 1964 and 1999, were constructed using ArcView 3.1.

Groundwater and surface water

Six locations on a ~60 m transect from the centre of the bog through the lagg and up the surrounding forested hillslope (F to B) were instrumented for measurements of water table position, hydraulic head, and water chemistry (Fig. 1). Piezometers made

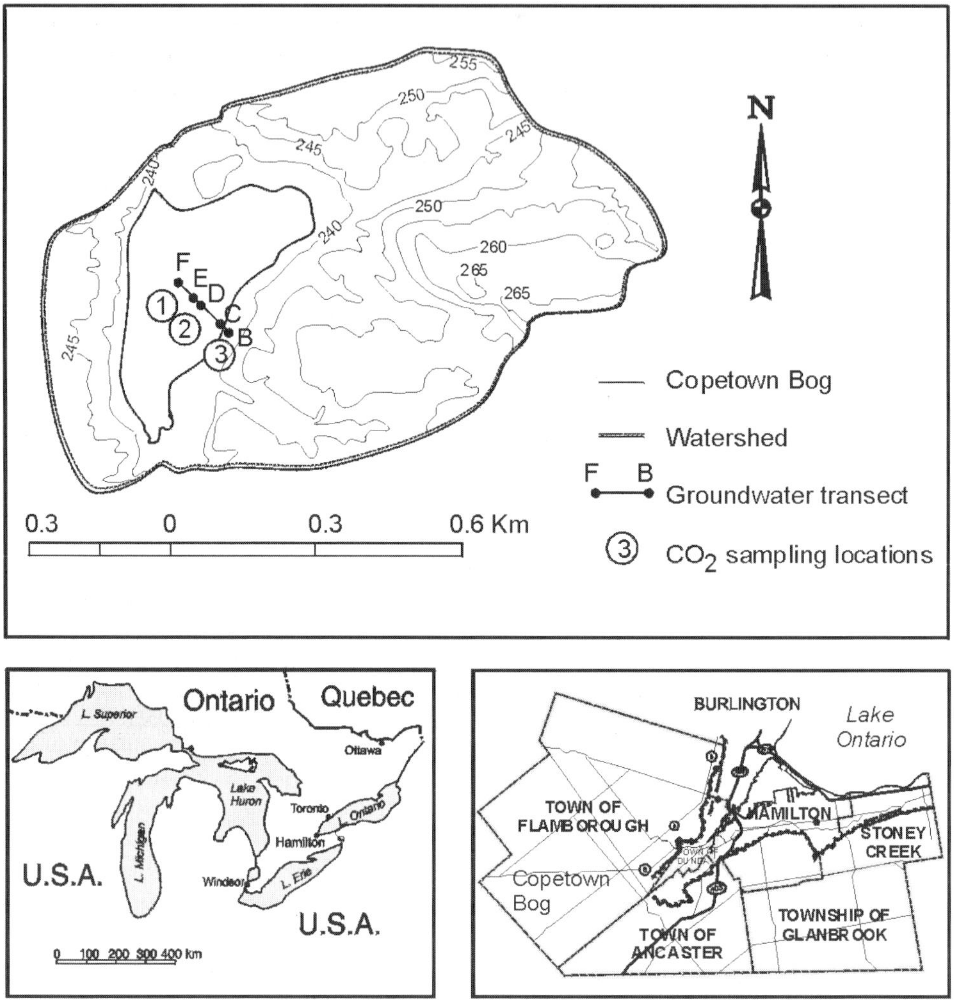

Fig. 1 Study site and sampling points at Copetown Bog. B–F indicate the groundwater piezometer nets and 1, 2 and 3 refer to the CO_2 sampling points (central, margin and lagg, respectively).

of 1.5-cm ID PVC pipe screened over the bottom 10 cm, were installed to depths of 0.5, 1.0, 1.5, 2.0 and 3.0 m. Groundwater flow nets and water chemistry were determined from bi-weekly measurements collected from September 1997 to July 1998. Surface water samples were collected from 11 locations within the lagg on four occasions from July to late November, 1998. All surface and groundwater chemistry samples were analysed for pH, conductivity, and NO_3^- and Cl^- using ion-specific electrodes.

Net CO_2 exchange

CO_2 exchange was measured over a one-week period each month using a Plexiglas climate controlled enclosure (surface area 0.25 m^2, volume 0.125 m^3) and an EGM-1 infrared gas analyser (PP Systems, UK) using standard techniques outlined in Waddington & Roulet (1996). Hourly measurements of air temperature and incident photosynthetically active radiation (PAR) at a meteorological site (Fig. 1), located between the central and margin sites, were utilized to model total net ecosystem CO_2 exchange (NEE) for the study period using a gross ecosystem production (GEP) and PAR relationship, and a total ecosystem respiration (R_{TOT}) and air temperature relationship (Waddington & Roulet, 1996).

RESULTS

Bog vegetation pattern change

Maximum vegetation change occurred in the lagg while the interior bog regions did not experience any change (Fig. 2). A total of nine species were present in 1999 that were absent in 1964, including: *Cornus stolonifera*, *Equisetum* sp., *Glyceria* sp., *Impatiens capensis*, *Quercus* sp., *Acer* sp., *Canibus* sp., *Poa* sp., *Saggittaria* sp., *Thelypteris palustris*. Three species: *Betula* sp., *Maianthemum canadense* and *Woodwardia virginica* disappeared completely between 1964 and 1999. The decrease in *Sphagnum* spp. (–8.8%) and *C. calyculata*, (–7.4%) and the appearance of *T. palustris* (+9.3%) characterize the major changes in the northern portion of the bog. *Impatiens capensis* appearance (+10% coverage) was concentrated in the southern portion of the lagg (Fig. 2(c)) and was not present in the centre of the bog. *Carex* sp. coverage increased by almost 13% of the bog area, with the majority occurring in the southwest portion of the bog (Fig. 2(d)).

Wetland hydrology

The lagg was always the wettest site within the bog and contained standing water (16.6 to 36.5 cm deep). Copetown Bog is likely located on a perched water table (Fig. 3) with groundwater flow vertically downward at the interior portion of the bog (Nest F, not shown in Fig. 3), but towards the hillside close to the lagg (Nests D, C and B). At this location the flow crosses the peat boundary into the silt layer. The water table gradient is unknown, as the position of the regional water table was not determined. No groundwater discharge from the watershed into the lagg occurred at our transect during our measurement period. However, there was visual evidence (high conductivity, no winter ice cover) of groundwater emergence "hot spots" in some zones within the lagg. These hot spots are likely the location of agricultural tile drains. Moreover, surface water inputs to the lagg during high intensity rain events in the spring and by snow melt runoff resulted in direct water input from the watershed.

Water chemistry

Groundwater NO_3^- concentration in the hillslope (0.6 to 6.0 mg l^{-1}, mean = 2.8 mg l^{-1}) was typically three times greater than in the bog peat (0.1 to 3.0 mg l^{-1}, mean =

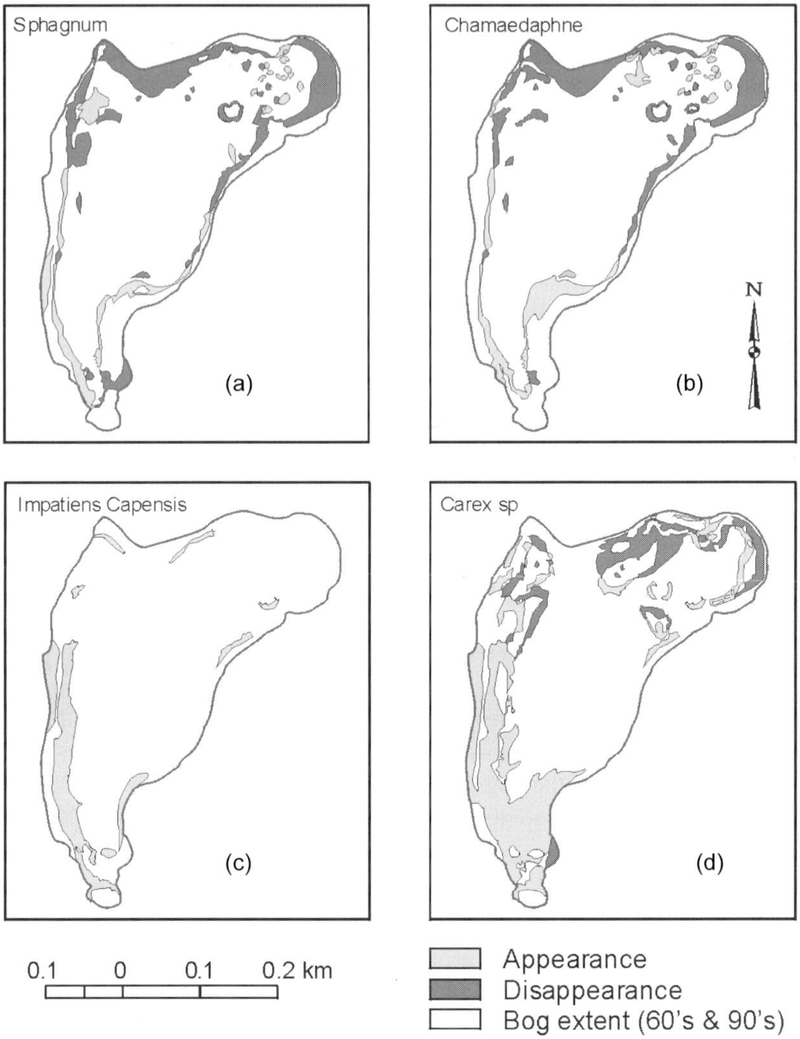

Fig 2 Appearance and disappearance of four key vegetation types at the Copetown Bog since 1964: (a) *Sphagnum* sp., (b) *Chamaedaphne calyculata*, (c) *Impatiens capensis*, and (d) *Carex* sp. Data are based on comparing our 1999 vegetation survey with Radforth (1964).

0.8 mg l^{-1}). The highest NO_3^- concentration in the silt was at 3.0 m depth, which may indicate a deeper groundwater source. There was also considerable variation in surface water chemistry in the lagg with NO_3^- concentrations ranging 1.3–27.6 mg l^{-1}. Electrical conductivity ranged from 25.7 to 300.7 μS cm^{-1} while pH ranged from 5.6 to 7.8.

CO_2 exchange

Seasonal GEP followed the trend: margin (–199.6 g C m^{-2}) > lagg (–140.4 g C m^{-2}) > central (–131.6 g C m^{-2}), while R_{TOT} followed the trend: lagg (273.3 g C m^{-2}) > margin

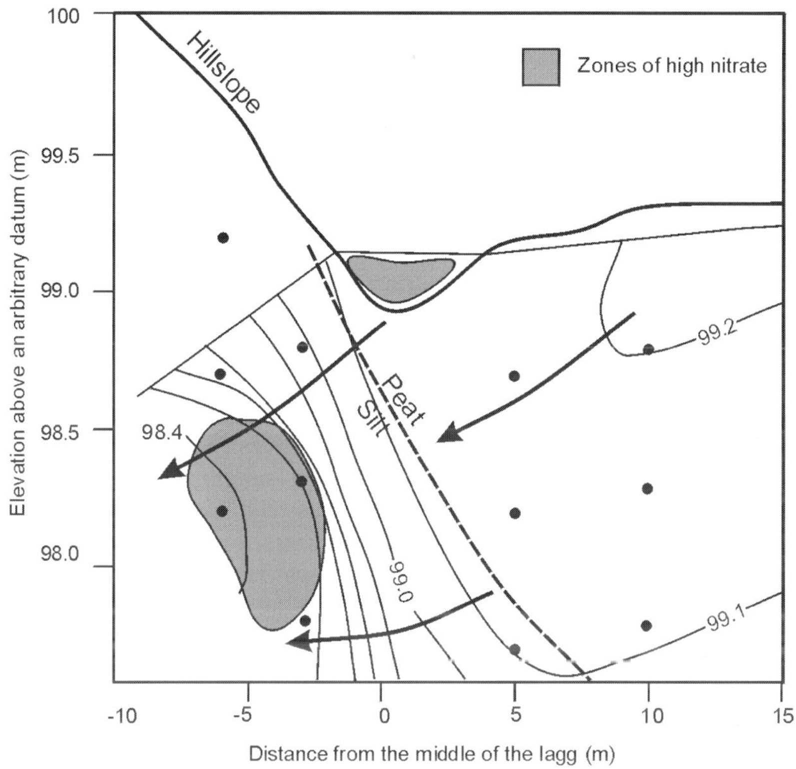

Fig. 3 Cross-section showing groundwater flow path at Copetown Bog showing nitrate movement through the subsurface along the transect, from the margin and across the lagg. Note: flow at the centre of the bog was vertically downward and is not shown.

(179.3 g C m^{-2}) > central (140.4 g C m^{-2}). Seasonal net ecosystem exchange of CO_2, determined by combining the seasonal GEP and R_{TOT} values, was 8.8, –20.3, and 132.9 g C m^{-2} at the central, margin, and lagg locations respectively. Using the results of the surface coverage (central = 27%, margin = 56%, lagg = 17%) the bog lost 13.6 g C m^{-2} during the study season.

DISCUSSION

Potential controls on changing vegetation patterns

Changing ecohydrological conditions (nutrient conditions and drying) are likely responsible for the vegetation pattern changes observed over the last 30 years at Copetown Bog. We suggest that vegetation has changed due to: (a) higher nutrient concentrations in, and adjacent to, the lagg, and (b) drier conditions at the southern end of the bog. Decreases in *Sphagnum* sp. is a key indicator that the conditions in Copetown Bog have become more nutrient rich since 1964 (Hodgetts, 1995). Nitrogen fertilization studies suggest an increase in carbon loss from peatlands due to depressed *Sphagnum* sp. production or enhanced decomposition (Gunnarsson & Rydin, 2000;

Aerts *et al.*; 1992). Vascular plants have root systems that allow nitrogen to be utilized via a completely different mechanism than bryophytes, which subsequently allows vascular plants to gain nitrogen from subsurface flow and not solely from precipitation, as is the case for bryophytes (Glaser *et al.*, 1990). Consequently, the increase in vascular plants at Copetown Bog, at the expense of *Sphagnum* sp., is an indicator of increased and also likely new sources of nitrogen available within the perimeter of the expanding lagg. Consistent with the observations of *Sphagnum* sp., the decrease in *Chamaedaphne calyculata* cover indicates that the northern portion of the bog has been influenced by nutrient additions from adjacent lands. *Chamaedaphne calyculata* is a small shrub commonly found in mature ombrotrophic bogs whose source of water is primarily direct precipitation (Tiner, 1999). *Impatiens capensis* is known to border bogs and peatlands as it is shade tolerant; however, typical lagg environments lack the nutrients and aeration necessary for these plants to survive. Zampella & Laidig (1997) identified this species as an indicator of disturbance when it appeared in acidic, nutrient-poor environments similar to Copetown Bog. The appearance of *Impatiens capensis* in the lagg suggests either enrichment favouring the growth of minerotrophic species, or a drying of the western portion of the lagg.

Current agriculture practices within the watershed of the Copetown Bog appear to influence NO_3^- levels in the lagg as the northeasterly site in the lagg (closest to most of the agricultural land) had the highest NO_3^- concentration in the bog (> 28 mg l^{-1}). However, hydrological data suggest that Copetown Bog is located on a perched water table and that the bog is not connected to local groundwater flow from the watershed. Instead NO_3^- is likely to be supplied to the bog lagg from water that enters the lagg through agricultural tile drains, culverts and overland flow (during snowmelt and spring-events). Copetown Bog, like most bogs, is raised at its centre, but the prominence of low-lying animal tracks has created a hydrological connection between the lagg and the central portion of the bog. During the period of study the physical signs of connection were present; however, the NO_3^- levels in the central portion of the bog suggest only a marginal influence on bog water chemistry. This is the first study of nitrate inputs to Copetown Bog and as a consequence we do not have direct evidence that nitrate levels have increased since 1964. Future studies should investigate peat degradation and vegetation pattern change in kettle-hole bogs that are not influenced by anthropogenic nutrient inputs to further assess this potential water quality impact on peatland ecosystems.

The changes observed in the plant populations in the southern portion of the bog suggest that conditions have become drier since 1964. This is supported by the loss of a significant portion of open water area in the southern lagg as well as colonization by species adapted to drier conditions (Fig. 2). This may be the result of a 1-m deep pond that was dug at the southern tip of the bog. The creation of this pond may have increased the hydraulic gradient in the southern portion of the bog since the pond is at a lower elevation than the adjacent lagg. However, we lack hydrological data to confirm this hypothesis.

A degrading peatland?

Previous paleoecological research in Copetown Bog indicates that the bog formed 9250 years BP, based on McAndrews (1976) analysis of a nearby kettle lake

(McHattie, 1992). Like many southern Ontario bog and lake sediments that show a marked rise in ambrosia (ragweed) pollen in 1850 (approximately 150 years BP; McAndrews, 1976), an ambrosia rise is also present in the upper 1 m of the Copetown Bog peat layer. This ambrosia rise has been interpreted as resulting from land clearing due to European settlement (Terresmae, 1972). Based on bulk density and peat depths, Copetown bog accumulated 36.6 g C m^{-2} $year^{-1}$ from peatland initiation to pre-European settlement. However, since European settlement, the carbon accumulation rate has decreased slightly to 31.7 g C m^{-2} $year^{-1}$. This contrasts with previous research (e.g. Warner *et al.*, 1989) and suggests that modern era agricultural disturbance (fertilization) may enhance lagg and bog margin decomposition. Our CO_2 flux measurements suggest the lagg region lost over 130 g C m^{-2} during the growing season and that, together with the expansion of the lagg, has led to peat degradation at the margins of the peatland and shifted the Copetown Bog to a net growing season CO_2 source (13.6 g C m^{-2}).

CONCLUSION

The shift in vegetation from mature bog species to those typical of a poor fen as well as high pore water NO_3^- concentrations and circum-neutral pH suggests that Copetown Bog may be reverting to a poor fen. The development of a raised or ombrotrophic bog occurs over a period of thousands of years; however, once formed, these systems demonstrate remarkable stability provided the peat accumulation exceeds decomposition (Clymo, 1984). While the natural transition from a bog to a poor fen may take several thousand years under natural conditions, Clymo (1984) suggests that human disturbances that alter nutrient dynamics, including fire, drainage and cultivation, may shorten this period to as little as 50 years. Cultural enrichment from agricultural practices could lead to similar outcomes provided hydrological connections are present. This appears to be the case for Copetown Bog. While the watershed is not connected to the bog via groundwater, management practices should be taken to remove the surface hydrological connections with the adjacent lands. Kettle-hole bogs are unique ecosystems and require protection to preserve their biodiversity.

Acknowledgements We wish to thank the Hamilton Conservation Authority for permission to use the Copetown Bog for this study. Thanks to S. Day, P. Tatham, M. Closson, K. Warner and J. Malloy for their field and laboratory assistance. B. Branfireun, J. Price, E. Kellner, B. Duncan, and M. Strack provided helpful comments during the preparation of this manuscript. The comments from an anonymous reviewer are also greatly appreciated.

REFERENCES

Aerts, R., Wallen, B. & Malmer, N. (1992) Growth-limiting nutrients in *Sphagnum*-dominated bogs subject to low and high atmospheric nitrogen supply. *J. Ecol.* **80**,131–140.

Bunting, M. J., Morgan, C. R., Van Bakel, M. & Warner, B. G. (1998) Pre-European settlement conditions and human disturbance of a coniferous swamp in southern Ontario. *Can. J. Bot.* **76**,1770–1778.

Clymo, R. S. (1984) The limits to peat bog growth. *Phil. Trans. Roy. Soc. London B* **303**, 605–654.

Glaser, P. H., Janssens, J. A. & Siegel, D. I. (1990) The response of vegetation to chemical and hydrological gradient in the Lost River peatland, northern Minnesota. *J. Ecol.* **78**,1021–1048.

Gunnarson, U. & Rydin, H. (2000) Nitrogen fertilization reduces *Sphagnum* production in bog communities. *New Phytol.* **147**, 527–537.

Hill, A. R. (1991) A ground water nitrogen budget for a headwater swamp in an area of permanent ground water discharge. *Biogeochem.* **14**, 209–224.

Hodgetts, N. (1995) Bog moss in Britain: the identification and role of *Sphagnum*. *Brit. Wildlife* **7**, 9–17.

McAndrews, J. H. (1976) Fossil history of Man's impact on the Canadian flora: an example from sounthern Ontario. *Can. Bot. Assoc. Bull.* **9**, 1–6.

McHattie, B. (1992) The Copetown Bog: a wetland evaluation. Hamilton-Wentworth Region Conservation Authority, Hamilton, Ontario, Canada.

Newmaster, S. G., Harris, A. G. & Kershaw, L. J. (1997) *Wetland Plants of Ontario*. Lone Pine Publishing and Queen's Printer for Ontario, Edmonton, Alberta, Canada.

Radforth, N. (1964) Summit Bog at Copetown, Ontario. In: *The Wood Duck*. Hamilton Field Naturalists Club Publication, Hamilton, Ontario, Canada.

Terasmae, J. (1972) Palynology and chemistry of lake sediment cores from southern Ontario, related to man's activities on the environment. *Research Report Series no. 10, Studies in Paleoecology no. 1*. Department of Geological Sciences, Brock University, St Catharines, Ontario, Canada.

Tiner, R. W. (1999) *Wetland Indicators: A Guide to Wetland Identification, Delineation, Classification and Mapping*. Lewis Publishers, London, UK.

Waddington, J. M. & Roulet, N. T. (1996) Atmosphere-wetland carbon exchanges: scale dependency of CO_2 and CH_4 exchange on the developmental topography of a peatland. *Global Biogeochem. Cycles* **10**, 233–245.

Warner, B. G., Kubiw, H. J. & Hanf, K. I. (1989) An anthropogenic cause for quaking mire formation in southwestern Ontario. *Nature* **340**, 381–384.

Zampella, R. A. & Laidig, K. J. (1997) Effect of watershed disturbance on Pinelands stream vegetation. *J. Tor. Bot. Soc.* **124**, 52–66.

Comparison of wetlands in different hydrogeological settings under conditions of extreme climate variability

THOMAS WINTER[1], DONALD ROSENBERRY[1], ERIN KELLY[1] & JAMES LABAUGH[2]

1 *US Geological Survey, MS 413 Denver Federal Center, Lakewood, Colorado, USA*
tcwinter@usgs.gov

2 *US Geological Survey, MS 411 National Center, Reston, Virginia, USA*

Abstract Wetlands in the Cottonwood Lake area in North Dakota, USA, are underlain by poorly permeable till and have little groundwater input. Lakes and wetlands in the Shingobee River headwaters in Minnesota are underlain by permeable sand and have substantial groundwater input. Hydrological, chemical, and biological characteristics of these ecosystems have been monitored since 1977. Both sites experienced the second worst drought of the 20th century followed by the wettest period in more than a century. At Cottonwood Lake, plants that invaded the dry wetlands during the drought were flooded during the wet period and became a food source for animals. This resulted in successive substantial population increases and declines of plankton, invertebrates, amphibians and waterfowl. Substantial groundwater input buffered the lakes and wetlands in the Shingobee area against the changing water conditions. Only subtle changes in water chemistry and plankton populations were observed during the transition from drought to deluge.

Key words climate change; groundwater; population dynamics; water budgets; water chemistry; USA; wetlands

INTRODUCTION

It has been documented for a number of hydrogeological settings that the chemistry of lake and wetland water is controlled to a large extent by the interaction of these surface water bodies with groundwater (LaBaugh, 1988; Sacks *et al.*, 1992; Krabbenhoft & Webster, 1995; Hayashi *et al.*, 1998; Winter, 2001). Less well documented are the structure and function of aquatic communities in surface water bodies that lie along groundwater flow paths (Euliss *et al.*, 2004). And, only rarely have chemical and biological responses been documented during a period of extreme climate variability in water bodies so situated along groundwater flow paths (LaBaugh & Swanson, 2003; Swanson *et al.*, 2003).

Two field sites were established in the north-central USA for the purpose of studying the ecology of isolated lakes and wetlands that were interacting with groundwater flow systems. The Cottonwood Lake area in North Dakota is underlain by low-permeability glacial till. The Shingobee headwaters area in Minnesota is underlain by permeable glacial outwash sand. Physical hydrology, water chemistry, and selected biological data have been collected since the sites were established in the late 1970s. During that time, the north-central USA was subjected to a severe drought that lasted

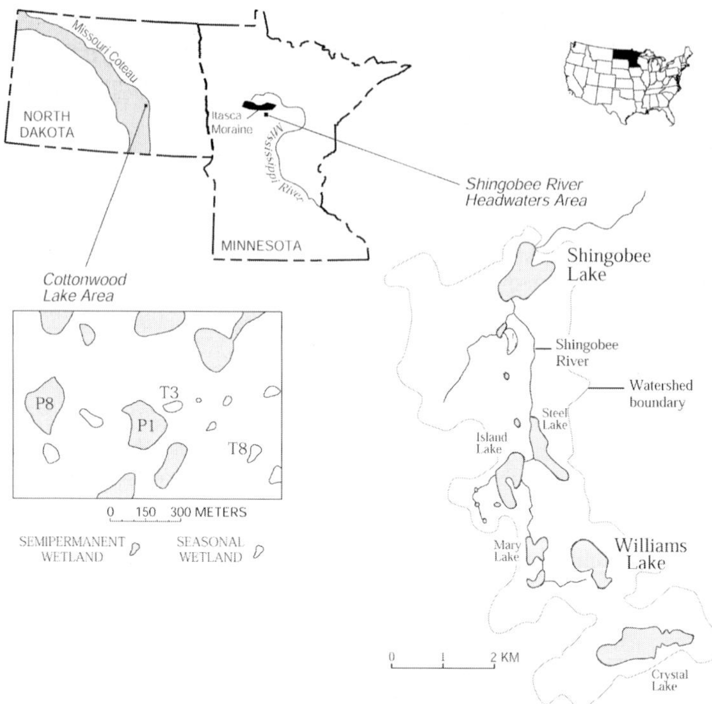

Fig. 1 Location of lakes and wetlands in the Cottonwood Lake area in North Dakota and in the Shingobee River headwaters area in Minnesota.

from 1988 to 1992. It was the second worst drought of the 20th century for this part of the USA. The drought was followed immediately by a deluge that began in 1993, and which caused high water levels that continue to the present. It was perhaps the wettest period in more than 150 years (Winter & Rosenberry, 1998). The purpose of this paper is to present the physical, chemical, and biological responses of the lakes and wetlands at these two sites to the extreme and sudden shift in climate from the drought to the deluge, and to relate the differences in response to exchange between groundwater and surface water.

DESCRIPTION OF FIELD SITES

The Cottonwood Lake area is situated high on the eastern edge of the Missouri Coteau, a large glacial moraine that transects North Dakota (Fig. 1). Local relief within the 80-ha Cottonwood Lake area is about 33 m. The area is very hummocky and has many small closed depressions, most of which contain wetlands. The glacial deposits in the Cottonwood Lake area are largely poorly permeable silty, clayey till (Winter & Rosenberry, 1998). Under normal conditions, wetlands in the Cottonwood Lake area are only a few ha in area and less than 2 m deep.

The Shingobee River headwaters area is located on a small morainal ridge that extends south from the much larger east–west trending Itasca Moraine (Fig. 1). Local relief in the area is about 67 m, and it is underlain largely by permeable sand. The area is characterized by hummocky topography containing numerous lakes and wetlands that interact with a regional groundwater flow system. Crystal Lake, which has no surface water inlets or outlets, lies near the drainage divide of both surface water and groundwater systems. Seepage from Crystal Lake moves through the groundwater system to Williams Lake, which also has no surface outlet. The Shingobee River begins as a seep in a wetland downgradient from Williams Lake. The river then flows through a series of lakes before entering Shingobee Lake. Research results are presented for two lakes in this paper: Williams and Shingobee. Williams Lake is 39 ha in area and 10 m deep. Shingobee Lake is 66 ha in area and 11 m deep.

The climate of the two sites is continental. Most air masses move from west to east, but polar air masses are common in winter and air masses from the Gulf of Mexico are common in summer. Annual precipitation averages about 44 cm in the Cottonwood Lake area and 66 cm in the Shingobee River headwaters area. Evaporation exceeds precipitation by about 30 cm in the Cottonwood Lake area but they are about equal in the Shingobee River headwaters area.

METHODS

At both sites, climate stations were established to continuously record precipitation and the variables needed to calculate evaporation by the energy-budget method (Sturrock *et al.*, 1992; Parkhurst *et al.*, 1998). Lake and wetland stages were continuously recorded or staff gauges were read daily. Stream discharge to and from Shingobee Lake was measured by continuously recording stage and developing a stage–discharge relationship. More than 60 wells were installed throughout each study area to determine the interaction of the lakes and wetlands with groundwater flow systems. The wells consisted of PVC pipe and screens. They were completed such that the screened interval was just below the water table.

Chemical analysis of surface and groundwater from the study sites included pH, specific conductance, Ca, Mg, Na, K, Cl, SO_4, alkalinity, Si, Fe, Mn, soluble reactive phosphorus, total phosphorus, NO_3, ammonia, and total Kjeldahl nitrogen. Chlorophyll-*a* also was determined. Water from the wetlands at the Cottonwood Lake area was collected for chemical analysis on a biweekly to monthly basis when the wetlands contained water (LaBaugh *et al.*, 1996). Water from the Williams and Shingobee lakes in the Shingobee River headwaters area was collected for chemical analysis on a biweekly to monthly basis, and on occasion from other lakes and wetlands in the watershed (LaBaugh, 1988; LaBaugh *et al.*, 1995). Groundwater was collected for chemical analysis from wells at both sites on an irregular basis. Groundwater was also collected repeatedly in months between April and November from selected wells at the Cottonwood Lake area between 1980 and 1982, and in the vicinity of Williams Lake between 1981 and 1992 (LaBaugh *et al.*, 1995, 1996). Further details about sample collection, collection frequency, and laboratory analyses are provided in LaBaugh *et al.* (1995, 1996).

Vegetation and vegetation zones in selected wetlands at the Cottonwood Lake area were recorded by aerial photography and by plant identification along randomly

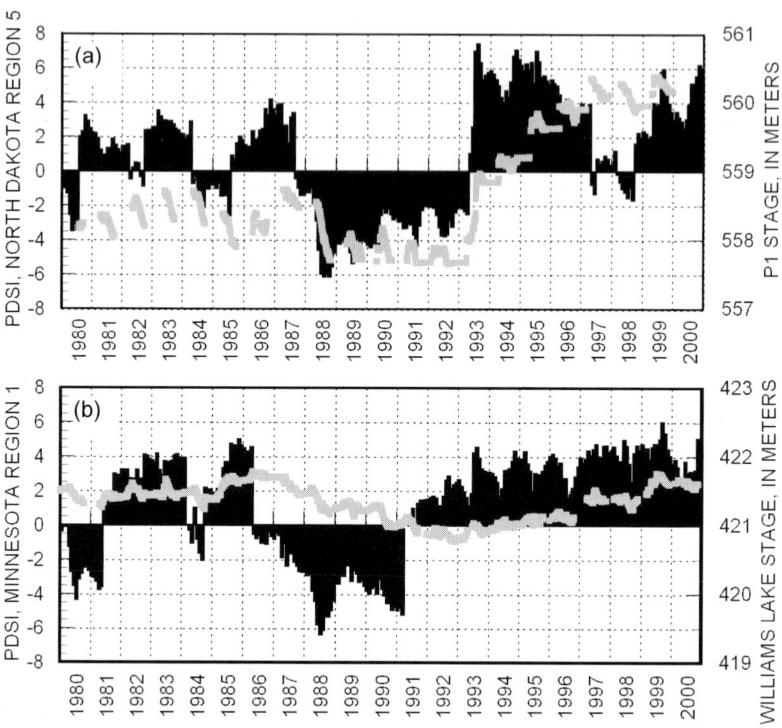

Fig. 2 Relation of the Palmer Drought Severity Index to: (a) water levels of Wetland P1 in the Cottonwood Lake area, and (b) water levels of Williams Lake in the Shingobee River headwaters area. Black bars are PDSI values; grey lines are wetland or lake stage.

selected transects as described by Swanson *et al.* (2003). Vegetation within selected lakes at the Shingobee headwaters area site was also examined using methods described by Carter *et al.* (1997). Algae and invertebrates in the water column of the lakes and wetlands were collected and identified by methods described by LaBaugh & Swanson (2003).

RESULTS

Climate variability and water levels

From the beginning of these studies in the late 1970s until 1987, the climate at both sites showed minor variations between wet and dry years (Fig. 2). This variability was reflected in the water level fluctuations of the lakes and wetlands. The extremely dry conditions in 1988 caused water levels to decline sharply in North Dakota, and they remained low for the next five years (Fig. 2(a)). The deluge that began in 1993 caused water levels to rise sharply and they have remained high to the present. In Minnesota, water level declines were gradual over the period of the drought, and their rise also was gradual during the wet period (Fig. 2(b)).

Cottonwood Lake area, North Dakota

Wetlands in the Cottonwood Lake area have relatively small surface watersheds and are underlain by till having low permeability. Hydraulic conductivity (*K*) of the unweathered till is about 10^{-8} cm s^{-1} and *K* of the fractured till near the land surface is about 10^{-4} to 10^{-7} cm s^{-1}. As a result, groundwater is a minor contributor; most of the water supplied to the wetlands (90% or more) is from precipitation. Their strong dependence on precipitation is reflected in the close correlation between wetland water levels and the Palmer Drought Severity Index (Fig. 2(a)). During the drought, all of the wetlands became completely dry, although they would hold small volumes of water for a week or two following summer rainfall events. During the deluge, water levels rose so high that individual wetlands merged to become open-water lakes (Winter & Rosenberry, 1998). Wetlands normally 1 m deep became as much as 4 m deep. Seasonal wetlands that normally held a fraction of a metre of water for a few weeks to about two months in spring held water throughout the summer.

Although groundwater is a minor contributor of water to these wetlands (10% or less) of the total input, it is the major contributor of solutes (LaBaugh, 1988). As a result, the water chemistry of the wetlands reflects their relation to groundwater. Wetlands with the greatest freshwater volume are generally in the recharge areas of groundwater flow systems. The most saline wetlands are in discharge areas of regional flow systems. Wetlands having intermediate levels of salinity usually have seepage from and seepage to groundwater. In addition to salinity differences, wetlands located in different parts of the groundwater flow field can have different major ion water types (Fig. 3(a)). It is noteworthy that during the great changes in water volume between normal, dry, and deluge conditions, major ion water type changed in some of the wetlands. For example, Wetland P1, which commonly contains sulfate water, became a bicarbonate system as a result of the deluge. Concentrations of sulfate in the water of Wetland P1 varied between 11.2 and 104 meq l^{-1} from 1984 to 1992 (143 analyses); whereas, in 1993 concentrations of sulfate varied between 3.3 and 12.6 meq l^{-1} (39 analyses). In the same time periods, alkalinity concentrations varied between 1.8 and 11 meq l^{-1} (1984-1992) and between 1.5 and 7.3 meq l^{-1} (1993). After mid August of 1993 and throughout 1994, alkalinity concentrations exceeded sulfate concentrations, although the range of values was similar, 5.3 to 8.2 meq l^{-1} for alkalinity and 3.3 to 7.2 meq l^{-1} for sulfate (17 analyses). Wetland P8, which normally contains bicarbonate water, became a sulfate system during the drought (LaBaugh *et al.*, 1996). Concentrations of sulfate in the water of Wetland P8 varied between 1.2 and 5.4 meq l^{-1} from 1984 to 1987 (77 analyses). During the drought (1988–1992), sulfate concentrations varied between 3.9 and 71 meq l^{-1} (77 analyses). In the same time periods, alkalinity concentrations of Wetland P8 water varied between 5.7 and 9.6 meq l^{-1} (1984–1987) (75 analyses), and between 2.5 and 12 meq l^{-1} (1988–1992) (77 analyses). In Wetland P8, the years of rising water levels (1993–1994) were ones in which concentrations of sulfate varied between 3.1 and 6.6 meq l^{-1} (43 analyses), while alkalinity concentrations varied between 1.6 and 7.7 meq l^{-1}.

During the relatively normal years before the drought, the distribution of wetland vegetation was fairly typical of prairie pothole wetlands. Generally, concentric bands of deep marsh, shallow marsh, and wet meadow progressed outward from a central

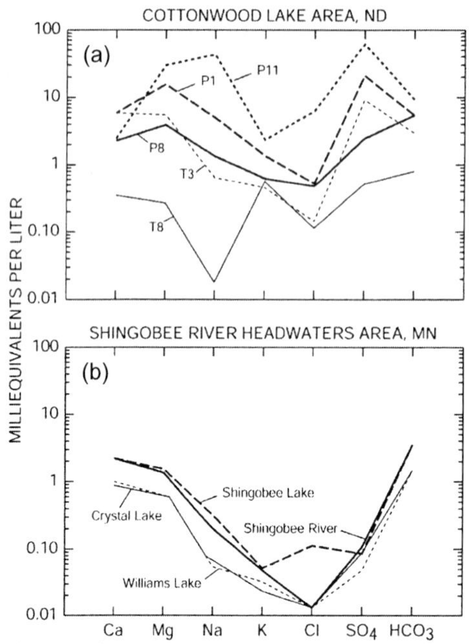

Fig. 3 Major-ion water types in: (a) wetlands in the Cottonwood Lake area (Wetland P11 is 3 km west of the study area), and (b) lakes in the Shingobee River headwaters area.

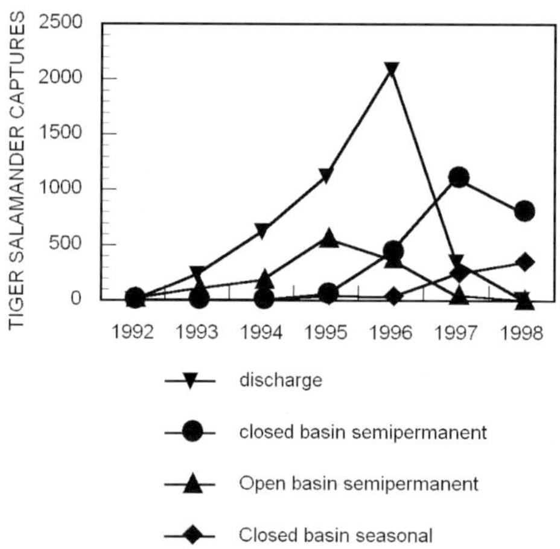

Fig. 4 Number of tiger salamander captures for various wetland types for the first five wet years following the drought of 1988-1992.

open water area. During the drought, cattail, which generally occur in the deep marsh zone of semipermanent wetlands, gradually invaded into the central open water areas. By the fifth year of the drought, monotypic stands of cattail covered the central areas of semipermanent wetlands, and terrestrial plants moved into the small seasonal wetlands (Swanson *et al.*, 2003). The sharp rise in water levels caused by the deluge, and the persistent high stages resulted in the death of all of these plants by drowning. This vegetative biomass became a food source for lower animal organisms, which in turn became a food source for higher animals. Thus, animals that live in the wetlands, such as zooplankton and invertebrates, and animals that use the wetlands, such as amphibians and waterfowl, had population explosions that were followed by population declines as the food sources were depleted. Interestingly, the peak populations of different organisms occurred at different times in different wetland types (Euliss *et al.*, 2004) (Fig. 4).

Shingobee River headwaters area, Minnesota

The glacial deposits are much more permeable in the Shingobee area than they are in the Cottonwood Lake area. Hydraulic conductivity of the sand ranges from about 10^{-3} to 10^{-5} cm s^{-1} As a result, the streams, lakes, and wetlands have substantial interaction with groundwater. Groundwater is the major source of water to Williams Lake and the major loss of water from the lake is seepage to groundwater (Rosenberry *et al.*, 1997). Streamflow to and from Shingobee Lake are the major components of its water budget, but groundwater contributes a substantial amount of inflow to the Shingobee River. Despite streamflow being the primary source of water to Shingobee Lake, groundwater still contributes about 25% of the inflow to the lake (Rosenberry *et al.*, 1997).

During the years prior to the drought, the stage of Williams Lake had minor fluctuations within a range of about half a metre (Fig. 2(b)). During the five years of drought, the stage of Williams Lake gradually declined 0.8 m. From the start of the wet period in 1993 to 2001, the lake stage increased by 1.2 m to the highest level the lake had reached during the study. Studies of Shingobee Lake only began during the drought, in 1989, but the lake stage declined 0.6 m by 1993. Because the lake has a surface outlet, its stage did not rise during subsequent years of the wet period.

The water chemistry of the lakes in the Shingobee River headwaters area also reflect their relation to groundwater (LaBaugh, 1988). The freshest lake, Crystal, is in the recharge area of the groundwater flow system and Shingobee Lake, the most mineralized of the group, is in the regional discharge area. However, unlike the conditions at Cottonwood Lake, the differences in chemistry among lake waters are far more subtle. For example, all of the lakes have calcium bicarbonate water types, and the salinity difference between the highest and lowest lakes is very small (Fig. 3(b)). The long-term trends in the chemistry of Williams and Shingobee lakes showed little variation related to changing climate or lake volumes. Calcium concentration in Williams Lake increased slightly, but not significantly, during the drought and remained at a higher concentration during the wet period, but there was essentially no change in calcium concentration in Shingobee Lake during the entire course of the study.

Unlike the aforementioned response of the biota at Cottonwood Lake to the extreme variations in climate, the response of the aquatic biota in the Shingobee area lakes was minimal. Studies of the biota in Williams and Shingobee Lakes were not as extensive and detailed as at Cottonwood Lake. Nevertheless, no changes in type or distribution of macrophytes were noticed in either lake. Extensive sampling of phytoplankton and zooplankton in the two lakes allowed careful examination of the response of these organisms to the shift from drought to deluge. The only noticeable difference in the phytoplankton communities was that the number of cyanophyte (blue-green algae) species in Williams Lake increased from 5 in the last dry year, 1992, to 9 in 1993 and 10 in 1994. Conversely, the number of cyanophyte species in Shingobee Lake decreased substantially in the transition from dry to wet years: 11 species were present in 1992, 5 in 1993, and 3 in 1994. There were no noticeable differences in the zooplankton communities of either lake between the dry and wet years.

DISCUSSION

The substantial difference in physical, chemical, and biological responses to extreme climate variability at the Cottonwood Lake site in North Dakota and at the Shingobee River headwaters area in Minnesota indicates the great influence groundwater has on these ecosystems. Wetlands in the Cottonwood Lake area derive 90% or more of their water supply from precipitation and lose 90% or more of their water to evapotranspiration. The small contribution from groundwater results in the water volume of these wetlands having little buffering capacity against variations in climate. On the other hand, the tremendous variations of water volume, water chemistry, and biological responses result in an extremely dynamic and productive ecosystem. These dynamic conditions are the foundation for the prairie pothole region being the primary waterfowl breeding ground for North America.

Groundwater has a substantial hydrological effect on lakes, wetlands, and streams in the Shingobee River headwaters area. It is the major source of water to the lakes having no stream inflow and it contributes large amounts of water to the Shingobee River. Although stream inflow is the largest source of water to Shingobee Lake, water in the stream is a combination of direct groundwater inflow and outflow from upstream lakes, which receive substantial amounts of groundwater. The size and permeability of the aquifer and groundwater flow system underlying the Shingobee area provides a very effective buffer against extreme variations in climate. Even though lake level declined by as much as 0.8 m over a 4-year period in Williams Lake, the water chemistry and plankton communities of the lake were not noticeably different during normal, dry, and wet conditions.

Palaeolimnological studies of the sediments of Williams and Shingobee lakes indicate that only major changes in climate, such as occurred following the ice age and during the postglacial hypsithermal interval, caused significant changes in the biota of the lakes (Schwalb & Dean, 2002). During the hypsithermal interval, for example, Williams Lake was 4 m lower than at present, which reduced its volume to about half of what it is today. Based on the significant changes in upland vegetation recorded in the lake pollen record from that time, it is likely that substantial within-lake processes also changed significantly in response to a climate change of that magnitude.

CONCLUSION

Results of studies of these two aquatic systems demonstrate the strong influence of groundwater on lake and wetland ecosystems. Lakes and wetlands underlain by permeable geological deposits and interacting with large groundwater flow systems are greatly buffered against even extreme variations in climate. As a result, changes in water levels are gradual and populations of aquatic organisms are little affected. In contrast, lakes and wetlands underlain by poorly permeable geological deposits are greatly affected by variations in climate. As a result, changes in water levels are large, water chemistry varies widely, and populations of aquatic organisms experience extreme variations in abundance.

Acknowledgement We thank personnel of the US Geological Survey's Northern Prairie Wildlife Research Center for collecting the field data at the Cottonwood Lake area in North Dakota.

REFERENCES

Carter, V. L., Rybicki, N. B., Striegl, R. G. & Gammon, P. T. (1997) Aquatic macrophytes in Williams and Shingobee Lakes – implications for carbon cycling. In: Hydrological and Biogeochemical Research in the Shingobee River Headwaters Area, North-Central Minnesota (ed. by T. C. Winter), 83–87. *US Geol. Survey Water Resources Investigations Report 96-4215*.

Euliss Jr, N. H., LaBaugh, J. W., Fredrickson, L. H., Mushet, D. M., Swanson, G. A., Winter, T. C., Rosenberry, D. O. & Nelson, R. D. (2004) The wetland continuum: a conceptual framework for interpreting biological studies. *Wetlands*, **24**(2), 448–458.

Hayashi, M., van der Kamp, G. & Rudolph, D. L. (1998) Water and solute transfer between a prairie wetland and adjacent uplands, 2. Chloride cycle. *J. Hydrol*. **207**, 56–67.

Krabbenhoft, D. P. & Webster, K. E. (1995) Transient hydrogeological controls on the chemistry of a seepage lake. *Water Resour. Res*. **31**, 2295–2305.

LaBaugh, J. W. (1988) Relation of hydrogeologic setting to chemical characteristics of selected lakes and wetlands within a climate gradient in the North-Central United States. *Verh. Internat. Verein. Limnol.* **23**, 131–137.

LaBaugh, J. W. & Swanson, G. A. (2003) Spatial and temporal variability in specific conductance and chemical characteristics of wetland water and in water column biota in the wetlands in the Cottonwood Lake area. In: Hydrological, Chemical, and Biological Characteristics of a Prairie Pothole Wetland Complex Under Highly Variable Climate Conditions—The Cottonwood Lake Area, East-central North Dakota (ed. by T. C. Winter), 35–53. *US Geol. Survey Prof. Paper 1675*.

LaBaugh, J. W., Rosenberry, D. O. & Winter, T. C. (1995) Groundwater contribution to the water and chemical budgets of Williams Lake, Minnesota, 1980-1991. *Can. J. Fish. & Aquat. Sci*. **52**, 754–767.

LaBaugh, J. W., Winter, T. C., Swanson, G. A., Rosenberry, D. O., Nelson, R. D. & Euliss Jr, N. H. (1996) Changes in atmospheric circulation patterns affect mid-continent wetlands sensitive to climate. *Limnol. Oceanogr.* **41**, 864–870.

Parkhurst, R. S., Winter, T. C., Rosenberry, D. O. & Sturrock, A. M. (1998) Evaporation from a small prairie wetland in the Cottonwood Lake area, North Dakota—an energy budget study. *Wetlands* **18**(2), 272–287.

Rosenberry, D. O., Winter, T. C., Merk, D. A., Leavesley, G. H. & Beaver, L. D. (1997) Hydrology of the Shingobee River headwaters area. In: Hydrological and Biogeochemical Research in the Shingobee River Headwaters Area, North-Central Minnesota (ed. by T. C. Winter), 19–23. *US Geol. Survey Water Resources Investigations Report 96-4215*.

Sacks, L. A., Herman, J. S., Konikow, L. F. & Vela, A. L. (1992) Seasonal dynamics of groundwater–lake interaction at Donana National Park, Spain. *J. Hydrol*. **136**, 123–154.

Schwalb, A. & Dean, W. E. (2002) Reconstruction of hydrological changes and response to effective moisture variations from north-central USA lake sediments. *Quaternary Sci. Rev.* **21**, 1541–1554.

Sturrock, A. M., Winter, T. C. & Rosenberry, D. O. (1992) Energy-budget evaporation from Williams Lake, a closed lake in north-central Minnesota. *Water Resour. Res.* **28**(6), 1605–1617.

Swanson, G. A., Euliss Jr, N. H., Hanson, B. A. & Mushet, D. M. (2003) Dynamics of a prairie pothole wetland complex: implications for wetland management. In: Hydrological, Chemical, and Biological Characteristics of a Prairie Pothole Wetland Complex Under Highly Variable Climate Conditions—The Cottonwood Lake Area, East-central North Dakota (ed. by T. C. Winter), 55–94. *US Geol. Survey Prof. Paper 1675*.

Winter, T. C. (2001) The concept of hydrologic landscapes. *J. Am. Water Resour. Assoc.* **37**(2), 335–349.

Winter, T. C. & Rosenberry, D. O. (1998) Hydrology of prairie pothole wetlands during drought and deluge: A 17-year study of the Cottonwood Lake wetland complex in North Dakota in the perspective of longer term measured and proxy hydrological records. *Climatic Change* **40**(2), 189–209.

4 Hydrochemistry and River Health

Suspended sediments in the Bermejo River and their impact on the hydrochemistry of the Lower Paraná River

L. DE CABO[1] & R. SEOANE[2]

1 *Museo Argentino de Ciencias Naturales "B. Rivadavia", Av. Angel Gallardo 470 (1405), Buenos Aires, Argentina*
ldecabo@macn.gov.ar

2 *Fac. de Ingeniería (UBA), Departamento de Hidraúlica, Av. Las Heras 2214 (1127), Buenos Aires, Argentina*

Abstract Based on data from our own and other sources, we estimated the numerical relationships between suspended sediments (SS) and the Bermejo River streamflow, identified the months when these relationships were statistically significant, and investigated the impact suspended sediments in the Bermejo River exert on the hydrochemistry of the Paraná River. Correlations were observed between suspended sediments and streamflows in the Bermejo River during the months of maximum and minimum streamflow. There were positive correlations between suspended sediments in the Bermejo and in the Middle Paraná River with a 1-month time lag with respect to the month of maximum streamflow; and between the Bermejo River streamflow and SS and calcium concentrations in the Lower Paraná River with a 2-month time lag. In the Lower Paraná River, the concentration of soluble reactive phosphorus was very low (5 μg l^{-1}) during the receding limb of the Bermejo River hydrograph and the rising limb of the Paraná River hydrograph; it was high (138–157 μg l^{-1}) during the Bermejo River high water season. The Bermejo River's huge yield (18 038 kg km^{-2} day^{-1}) plays an important role in SS, calcium and phosphorus concentrations in the Lower Paraná River.

Key words calcium; hydrograph; Paraná River basin; phosphorus; streamflow; suspended sediments; time lag

INTRODUCTION

Flowing waters constitute an important link in global biogeochemical cycles. These waters transport organic matter from terrestrial sources, generate organic material within aquatic environments, and degraded organic matter (Hedges *et al.*, 2000). In large rivers, these processes are influenced by the spatial and temporal availability of subsystems, such as flood plains, with a large water-holding capacity (Hein *et al.*, 2004). Biogeochemical transformations in flood plains are viewed as an important ecological service that these subsystems provide. Water quality preservation is dependent upon the interaction of hydrological patterns and the biogeochemical transformations that take place in the flood plains.

The Paraná River basin (2 600 000 km^2), which is part of the great Plata River basin (3 100 000 km^2), comprises regions with well-differentiated geological and climatic environments: the Brazilian shield and the Andes mountain range, both of which play an important role in the basin's sedimentology (Iriondo, 2004). The Bermejo River is the main tributary of the Lower Paraguay River that flows into the

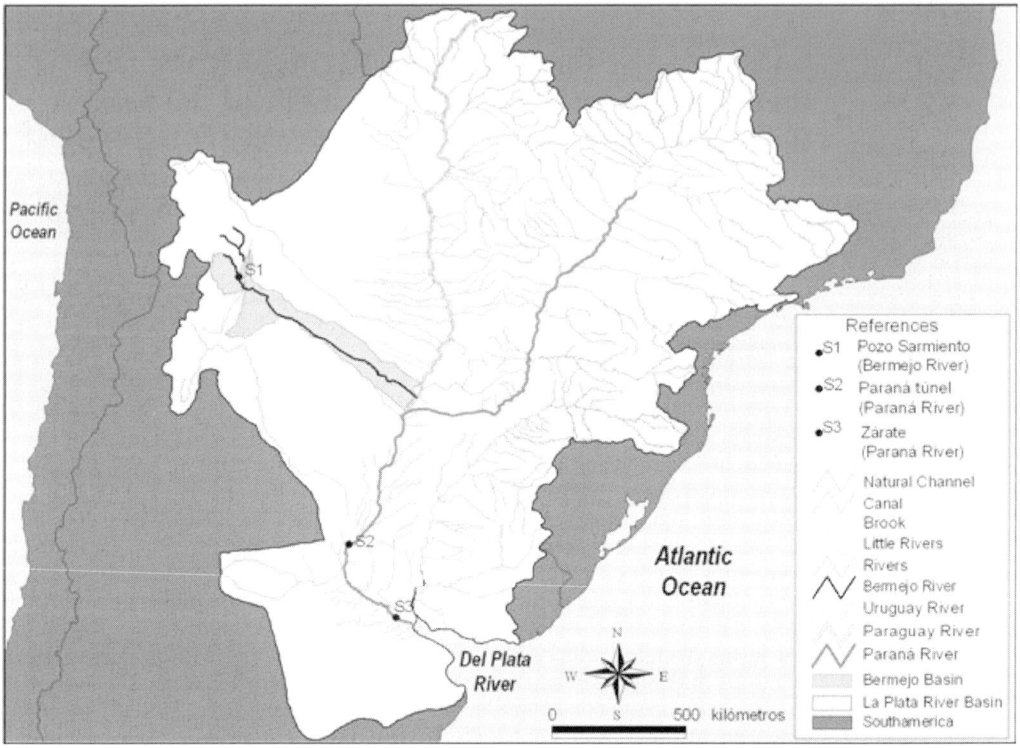

Fig. 1 La Plata River Basin.

Middle Paraná River (Fig. 1). The Bermejo River basin extends over 123 162 km^2 and its main watercourse is more than 1300 km long. Intensive erosion from a steep mountainous relief, little ground cover, and seasonal concentration of rainfall lead to high export rates (Pedrozo *et al.*, 1988).

Most of the water in the Paraná basin comes from the Upper Paraná River, while 50–70% of the total sediment load is supplied by the Upper Bermejo drainage basin (Depetris *et al.*, 2003).

This paper focuses on estimating a numerical relationship between suspended sediments (SS) and the Bermejo and Paraná river streamflows. It also seeks to identify the period of the year when these relationships are statistically significant. This aim allows the behaviour of the natural system before building the three dams projected for High Bermejo basin to be described. The impact of the Bermejo River on the concentration of suspended sediments, calcium (Ca) and phosphorus (P) in the Lower Paraná River is also assessed.

MATERIALS AND METHODS

EVARSA (Resource Assessment Corporation) furnished data on monthly streamflow data on the Bermejo River (at S1) and Paraná River (at S2) between 1906 and 2003 (Fig. 1), and on daily streamflows and SS concentrations at S1 and S2 for the 1993–2003 period. The quintile upper limits of monthly streamflow data at S1 and S2 were estimated. Records (97) of SS data from the Paraná River and 680 records of SS data for the Bermejo River were available for analysis. These discrete water samples were taken at 10 cm below the water surface and were collected from October 1990 to March 1995 in triplicate for chemical analyses from the navigation channel of the Lower Paraná River at S3 (Fig. 1). The water samples were prefiltered through Whatman GF/C glass-fibre filters (1.2 μm size particle retention). Then, the samples were filtered through 0.45-μm membrane filters and were placed in ice immediately after. Soluble reactive phosphorus (SRP) was determined with the ascorbic acid-phosphomolybdate method (Strickland & Parsons, 1972); calcium (Ca^{2+}) determinations were made by EDTA titration (APHA, 1985). Well-mixed samples were filtered through pre-weighed and dried Whatman GF/C glass-fibre filters, and the residue retained on the filters was dried to constant weight at 103°C. The increase in filter weight represented the total SS (APHA, 1985).

Correlations between SS and daily streamflow and between SS at various sites were performed. As the missing data had a random distribution, monthly correlation analyses were performed so as to prevent autocorrelation of time series data on suspended sediments and streamflows in the Bermejo and Middle Paraná rivers. A 5% level of significance was set for statistical analysis.

RESULTS AND DISCUSSION

Description of the study basins

The sub-basins of the Bermejo and Paraná rivers comprising the P. Sarmiento (S1) and Túnel (S2) stations are characterized in Table 1. S1 lies above sea level and annual rainfall, concentrated in storm events in the summer months, is lower than at S2, where rainfall distribution is more homogeneous year round. The Bermejo River has a single streamflow peak in February (summer), with 64% of streamflows in the first quarter

Table 1 Characterization of the study basins: General characteristics of database.

	Bermejo River (S1)	Paraná River (S2)
Station	P. Sarmiento	Túnel
Sub-basin area (km^2)	25 000	2 302 000
Latitude	23°06′	31°41′
Longitude	64°13′	60°30′
Altitude (station, metres above sea level)	296	9
Suspended sediment data period	1993–2003	1993–2003
Number of suspended sediments measurements	680	97
Hydrologic year	September–August	September–August
Streamflow data period	1940–2003	1904–2003
Month of streamflow peak	February (1027 m^3 s^{-1})	March (17417 m^3 s^{-1})
Month of minimum streamflow	September (49.24 m^3 s^{-1})	September (10795 m^3 s^{-1})
Month of highest SS conc. (mg l^{-1})	January (7795)	March (319)
Month of lowest SS conc. (mg l^{-1})	August (179)	August (104)
Month of highest sediment load (kg s^{-1})	March (10124)	April (6617)
Month of lowest sediment load (kg s^{-1})	August (16)	August (1573)
Mean annual rainfall (mm)	500	1100

of the year. At S2, the maximum and minimum streamflows of the Paraná River occur in March and September, respectively. However, at S2 streamflow distribution is more homogeneous.

According to the classification developed by Meybeck *et al.* (2003), the Bermejo River has a very high concentration of SS (8345 mg l^{-1}) and an extremely high yield (18 038 kg km^{-2} day^{-1}); the Paraná River, on the other hand, has a moderate concentration of SS (201 mg l^{-1}) and a moderate yield (123 kg km^{-2} day^{-1}) at S2. Compared to other rivers with high loads of suspended sediments (Meybeck *et al.*, 2003), the load in the Bermejo at S1 was moderate but, because of its smaller basin, its yield was highest due to very high erosion rates. Likewise, Brea & Spaletti (2002) studied the increase of SS load due to land cover changes in Bermejo basin.

Seasonal variations in the concentration of suspended sediments in the Bermejo River

High concentration of suspended sediments in the Bermejo River coincided with streamflow peaks, and low concentration with minimum streamflows (Fig. 2). Positive correlations between SS and streamflow in the Bermejo River were observed during months of minimum streamflow, August and May (Table 2), and months of maximum streamflow March and February (Table 2).

Impact of the Bermejo River on the Middle Paraná River

No significant correlation was found between SS and streamflow at S2. However, as shown in Table 1, minimum concentrations and loads of SS at S2 were recorded in

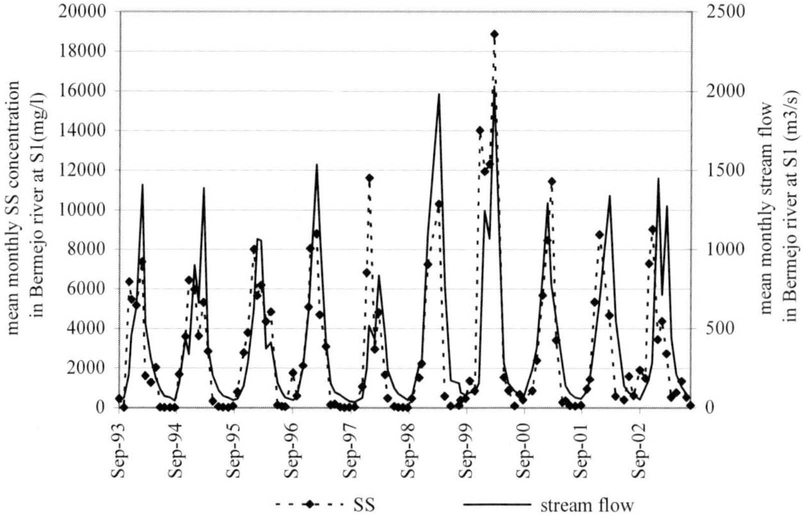

Fig. 2 Seasonal variation in mean monthly SS concentration and Bermejo River mean monthly streamflow at S1.

Table 2 Correlation coefficients between SS and streamflow in the Bermejo River (n = 10).

Month	r^2	p
September	0.081	0.823
October	0.261	0.461
November	0.160	0.659
December	–0.474	0.166
January	–0.28	0.434
February	0.703	0.035
March	0.636	0.048
April	–0.38	0.313
May	0.789	0.021
June	0.168	0.643
July	0.047	0.898
August	0.831	0.003

August, while maximum values were recorded in March and April, about a month after the streamflow peak. Drago (1990) reports that the concentration of SS in the Middle Paraná River lags 22 days behind streamflow peaks. Suspended solids at S2 and Bermejo River streamflow at S1 are related (Fig. 3). There was a significant correlation coefficient (r^2 = 0.711; n = 8) of SS concentration at S2 in the month of March (the month of maximum SS concentration at S2) and previous month SS concentration in the Bermejo River. Transport of sediments from S1 to S2 lagged about one month behind, suggesting that the Bermejo River is a significant source of sediment for the Paraná River with one month lag.

Impact of the Bermejo River on the Lower Paraná River

Bonetto *et al.* (1994) report that sediments in the Lower Paraná River are similar to those in the Bermejo River most of the year. Particle size analysis of SS in the Lower

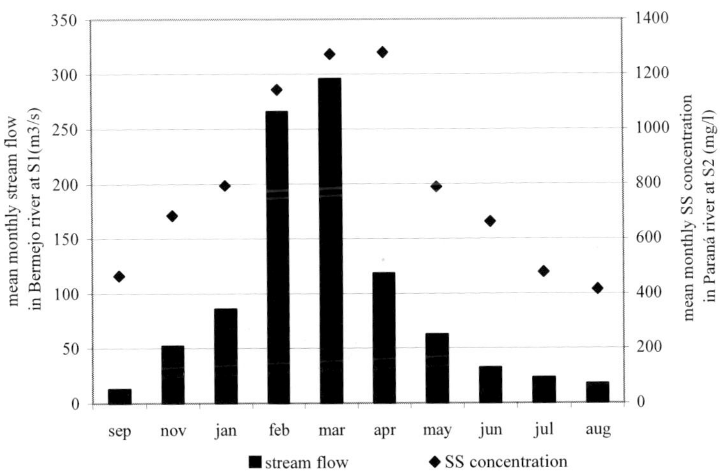

Fig. 3 Seasonal variation in SS concentration in the Middle Paraná River at S2 and Bermejo River mean monthly streamflow at S1.

Paraná River shows that they are mostly made up of silt-sized particles and aggregates (Iriondo, 2004), and of silt-sized particles in the Bermejo River (62%) (Bonetto & Orfeo, 1984). We found a significant correlation between SS concentration in the Lower Paraná River (S3) and the Bermejo River previous month streamflow at S1 ($r = 0.697$; $n = 13$; $P < 0.05$) (Fig. 4(a)). Therefore, both for the Bermejo River (range: 14–434 $m^3 s^{-1}$, mean: 126.14 $m^3 s^{-1}$) and the Lower Paraná River (range: 13578–23155 $m^3 s^{-1}$, mean: 18164 $m^3 s^{-1}$) streamflows, transport of suspended sediments from S1 to S3 would take approximately 1 month.

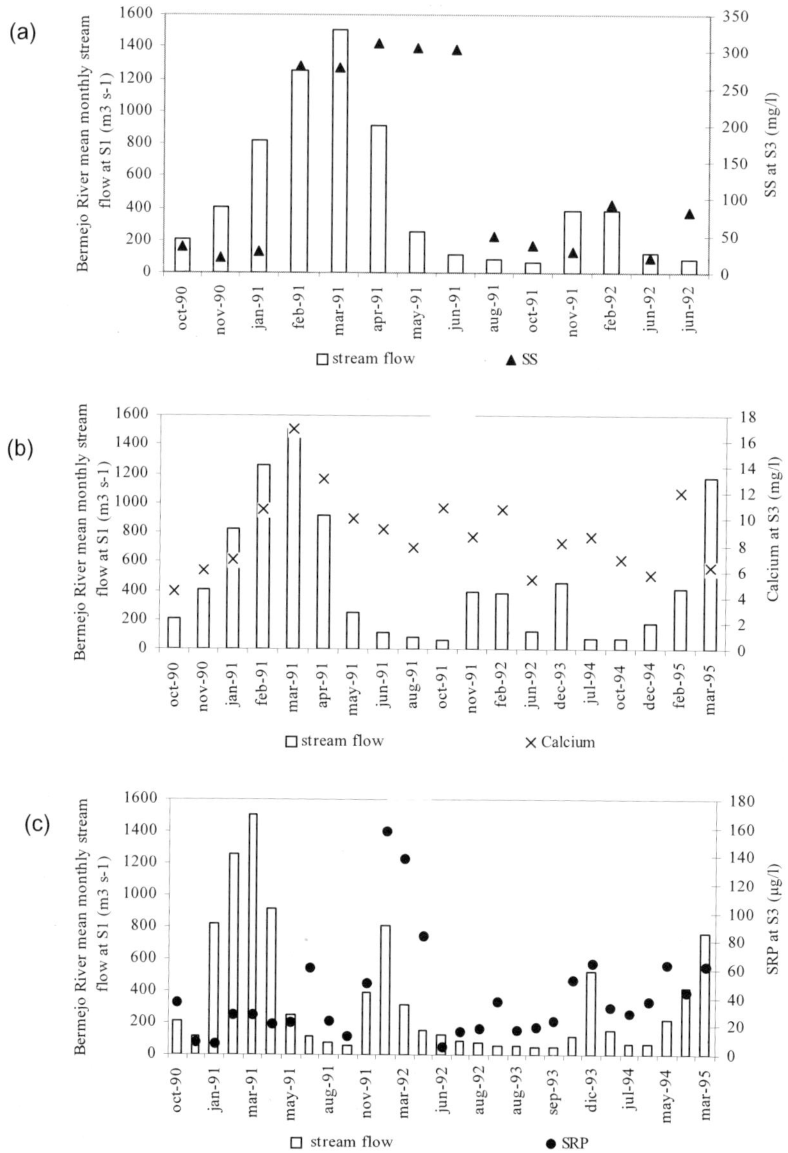

Fig. 4 Seasonal variation in: (a) SS, (b) calcium, and (c) SRP concentration in the Lower Paraná River at S3 and Bermejo River mean monthly streamflow at S1 ($m^3 s^{-1}$).

The significant correlation coefficient of Ca concentration at S3 and the Bermejo River streamflows, 2 months prior to sampling ($r = 0.769$; $n = 19$; $P < 0.05$) (Fig. 4(b)), showed that, on average, there was a 2-month lag between peak concentration of SS transported by the Bermejo River represented by streamflow and Ca concentration at S3. This may be due to the fact that the SS transported by the river are carbonate and Ca rich (Pedrozo & Bonetto, 1989). The considerable reduction in SS concentration in the Lower Paraná River compared with the values obtained by Drago & Amsler (1998) in the Middle Paraná River shows that sediment retention in the alluvial valley of the Lower Paraná River was high. Sediments are dissolved by the acid environment of the alluvial valley, increasing calcium and bicarbonate content in the water (Villar *et al.*, 1998) that flows into the main channel.

The average streamflow at S2 in the Paraná River in June 1992 (28 458 $m^3 s^{-1}$) was the highest of the time series for that month. At S2 the streamflow exceeded the 90% quintile for 230 days, coinciding with the 1991–1992 ENSO event. In June 1992, SRP values were very low (5 $\mu g\ l^{-1}$), coinciding with the winter high-water event of the Paraná at S2 that exerted a diluting effect following low streamflows in the Bermejo River the previous month (May) (ranking in the 50th–75th percentile). Another extreme situation was recorded in February–March 1992, when high levels of SRP (138–157 $\mu g\ l^{-1}$) were found in the Bermejo River during the high-water period. Villar *et al.* (1998) presented evidence that suggests the downstream SRP increase along the Lower Paraná River mainly originated from the water exchange between the river and its flood plain; the cultural point sources of SRP remain uncertain. Mean streamflow of the Bermejo River during the previous month (January) was in the 75th–100th percentile, while the streamflow of the Paraná River at S2 in February and March was in the 25th–50th percentile (Fig. 4(c)). Soluble Reactive P behaviour with respect to the Bermejo River streamflow may be explained by the fact that the Bermejo River sediments are transported by the Paraná River and deposited in its alluvial valley. The more acidic and occasionally anoxic environment in the valley favours the release of SRP into the water column, as demonstrated by sediment and water incubation experiments (de Cabo, 2001) and *in situ* tests (Villar *et al.*, 1998). When the SRP load from flood plain reached the main channel of the Paraná River, the SRP concentration in the main channel was increased; this process is associated with flow increasing.

Despite the distance between S1 and S3 (1700 km approx.), SS in the Bermejo River had a marked impact on the hydrochemistry of the Lower Paraná River. Suspended sediments, P and Ca in the Lower Paraná River appeared to be influenced by the Bermejo River streamflow. After the Bermejo River high-water event, with almost a 1-month lag, increased concentration of SS in the Lower Paraná River led to higher concentrations of P and Ca. If reservoirs were to be built across the Bermejo River, it is important that the load of SS should be maintained as the SRP and calcium concentrations in the Lower Paraná River-valley system depend on them.

Acknowledgements This research was possible thanks to the financial support of the International Foundation for Science (IFS). EVARSA (Resource Assessment Corporation) provided suspended solids and streamflow measurements. Prof. Priva Braunfeld made valuable comments on the manuscript.

REFERENCES

APHA (1985) *Standard Methods for the Examination of Water and Wastewater*, 14th edn. (ed. by American Public Health Association), Washington DC, USA.

Bonetto, A. & Orfeo, O. (1984) Caracteres sedimentológicos de la carga en suspensión del Río Paraná entre Confluencia y Esquina (Prov. de Corrientes, R. A.). *Rev. De la Asoc. Argentina e Mineralogía, Petrología y Sedimentología.* **15**(3–4), 51–61.

Bonetto, C., de Cabo, L., Gabellone, N., Vinocur, A., Donadelli, J. & Unrein, F. (1994) Nutrient dynamics in the deltaic flood plain of the Lower Paraná River. *Arch. Hydrobiol.* **131**(3), 277–295.

Brea, J. D. & Spaletti, P. D. (2002) Producción de sedimentos por erosión superficial, aplicación a Alta Cuenca del Río Bermejo y evaluación de eventuales cambios en la cobertura vegetal. *5º Encontro Nacional de Engenharia de Sedimentos. Asociación Brasilera de Recursos Hídricos,* San Pablo, Brazil.

de Cabo, L. (2001) Dinámica de nutrientes en el valle aluvial del Río Paraná Inferior. PhD Thesis, Universidad de Buenos Aires, Buenos Aires, Argentina.

Depetris, P. J., Probst, J. L., Pasquini, A. & Gaiero, D. (2003) The geochemical characteristics of the Paraná River suspended sediment load: an initial assessment. *Hydrol. Process.* **17**, 1267–1277.

Drago, E. C. & Amsler, M. L. (1988) Suspended sediment at a cross section of the Middle Paraná River: concentration, granulometric and inflence of the main tributaries. In: *Sediment Budgets* (ed. by M. P. Bordas & D. E. Walling) (Proc. Porto Alegre Symp., December 1988), 381–396. IAHS Publ. 174. IAHS Press, Wallingford, UK.

Drago, E. C. & Amsler, M. L. (1998) Bed sediment characteristics in the Paraná and Paraguay Rivers. *Water Int.* **23**, 174–183.

Drago, E. C. E. (1990) Geomorphology of large alluvial rivers: Lower Paraguay and Middle Paraná. *Interciencia* **15**(6), 378–387.

Hedges, J. I., Mayorga, E., Tsamakis, E., McClain, M. E., Aufdenkampe, A., Quay, P., Richey, J. E., Benner, R., Opsahl, S., Black, B., Pimentel, T., Quintanilla, J. & Maurice, L. (2000) Organic matter in Bolivian tributaries of the Amazon River: a comparison to the lower mainstream. *Limnol. Oceanogr.* **45**, 1449–1466.

Hein, T., Baranyi, C., Reckendorfer, W. & Schiemer, F. (2004) The impact of surface water exchange on the nutrient and particle dynamics in side-arms along the River Danube, Austria. *Sci. Total Environ.* **328**(1–3), 207–218.

Iriondo, M. (2004) The littoral complex at the Paraná mouth. *Quat. Int.* **114**, 143–154.

Meybeck, M., Laroche, L., Dürr, H. H. & Syvitski, J. P. M. (2003) Global variability of daily total suspended solids and their fluxes in rivers. *Global Planetary Change* **39**, 65–93.

Pedrozo, F. & Bonetto, C. (1989) Influence of river regulation on nitrogen and phosphorus mass transport in a large South American river. *Regulated Rivers: Res. Manage.* **4**, 59–70.

Pedrozo, F., Bonetto, C. A. & Zalocar, Y. (1988) A comparative study on phosphorus and nitrogen transport in the Paraná, Paraguay and Bermejo rivers. In: *Limnologia e Manejo de Represas* (ed. by J. G. Tundisi). Serie: Monografias em Limnologia **1**(1), 93–117, EESC-USP/CRHEA/ACIESP, Sao Carlos, Brazil.

Strickland, J. & Parsons, T. (1972) A practical handbook for seawater analysis. *J. Fish. Res. Bd. Can. Bull.* **167**, 311.

Villar, C., de Cabo, L., Vaithiyanathan, P. & Bonetto, C. (1998) River-flood plain interactions: nutrient concentrations in the Lower Paraná River. *Arch. Hydrobiol.* **142**, 433–450.

Revisiting biogeochemical aspects of the Paraná River

PEDRO J. DEPETRIS
Centro de Investigaciones Geoquímicas y de Procesos de la Superficie (CIGeS), FCEFyN, Universidad Nacional de Córdoba, Avenida Vélez Sarsfield 1611, X5016GCA Córdoba, Argentina
pdepetris@com.uncor.edu

Abstract The reassessment of biogeochemical data originally collected in the Paraná River between 1981 and 1984 highlights the role played by the long (~1200 km) and wide (~30 km) flood plain during the 1982–1983 ENSO-triggered exceptional flood. DOC is the only major biogeochemical variable that showed high (early flood stages) and low (late flood stages) concentrations during very high discharges (>22000 $m^3 s^{-1}$). Other variables, such as PCO_2, DIC, and the POC : PN ratio appeared to be more markedly affected by materials stored in the flood plain during the periods between successive extraordinary floods (the ENSO recurrence interval is 2–4 years), which are flushed out when such extreme events occur. TSS concentrations also decreased markedly during the high flood mainly because the over bank stage and lower water surface slopes implied lower water velocities and, hence, diminished sediment carrying capacity.

Key words biogeochemistry; carbon; dissolved organic matter; ENSO; flooding; fluvial dynamics; Paraná River; particulate organic matter

INTRODUCTION

The biogeochemical characteristics of rivers and wetlands have been the object of earnest scientific attention during the last 20 years. It was initially considered important to probe into the transfer rate of organic matter from continents to the World's oceans (e.g. Committee on Flux of Organic Carbon to the Ocean, 1981), but then it became apparent that it was also important to understand better the biogeochemical processes that occur in the rivers themselves (e.g. Richey *et al.*, 1990; Degens *et al.*, 1991).

It is known that rivers transport labile (i.e. consumable) or refractory organic matter, which may be allochthonous—if it is land-derived plant or animal debris, or if it is supplied by soil erosion, or autochthonous (i.e. the result of aquatic biological productivity). In large river systems located on continental passive margins, the extensive flood plains become additional sources of organic material in-as-much as the ox-bows, ponds, or abandoned channels that are close to the main waterway are annually inundated. Less frequently, due to exceptional flood events, water enters remote water bodies that have been disconnected from the main stem during several years, and covers islands and the emerged land up to the valley's edge. Hence, the flood periodically transfers partially-processed dissolved and particulate material to the river's main stem and thus it greatly affects the riverine organic budget. In this contribution, the biogeochemical processes that the Paraná River—as an example of other large systems—exhibits during extreme flooding is reviewed, placing emphasis on the role played by the peculiar hydrological conditions that prevail during such floods. This would assist in our understanding of the response of large fluvial systems to the increased runoff that global climate change may trigger in some cases.

HYDROLOGICAL FRAMEWORK

The Paraná River (~15° to ~35°S, ~50° to ~60°W) is one of the largest fluvial systems in South America (Fig. 1). It is over 4000 km long, its drainage basin has an area of 2.6 × 10^6 km^2 and its current long-term (1904–2003) mean discharge at Corrientes, Argentina, (about 1200 km upstream from the mouth; ~70% of the total drainage basin area) is about 16 200 m^3 s^{-1}. Paraná's middle and lower stretches are occupied by a long (~1200 km) and wide (~30 km) flood valley. The tropical Upper Paraná—with

Fig. 1 Map of the Río de la Plata drainage basin, with its two main tributaries, the Paraná and the Uruguay rivers. Total drainage area is 3.1 × 10^6 km^2 of which ~84% corresponds to the Paraná River basin.

headwaters in Brazil's Serra dos Preneos—contributes about 80% of the total discharge during high flow (January–April), which occurs during summer (in the southern hemisphere), and about 65% during the low-discharge period (June–August). The remainder is mainly supplied by the Paraguay River, which drains Brazil's Mato Grosso and the Gran Pantanal as well as the Andean eastern slopes through the Bermejo and Pilcomayo rivers (Fig. 1). The Paraná discharge series exhibits a significant linkage with the El Niño–Southern Oscillation (ENSO) events in the Equatorial Pacific, which has been underlined by several authors (e.g. Kousky *et al.*, 1984; Ropelewski & Halpert, 1987; Depetris *et al.*, 1996; Amarasekera *et al.*, 1997). The ENSO-related precipitation in the Paraná River drainage basin is in phase with the normal annual precipitation cycle, thus causing exceptional and disastrous flooding events during El Niño years. Besides the significant ENSO signal in the Paraná, Robertson & Mechoso (1998) also identified a near-decadal cycle that is associated with sea-surface temperature (SST) anomalies over the northern tropical Atlantic Ocean, and a 15- to 18-year cycle that appears to be associated with changes in the intensity of the South American summer monsoon and the southward low-level jet transport of moisture along the eastern slopes of the Andean chain (Robertson & Mechoso, 2000). Moreover, the Paraná River—and other rivers in southeastern South America—seems to be responsive to global change inasmuch as it is also showing a significant nonlinear increasing trend in the available 100-year-old discharge series, which apparently started in the 1970s (Genta *et al.*, 1998).

ORGANIC MATTER SOURCES

Based on the analyses of dissolved and particulate carbohydrates and amino acids, earlier research (Depetris & Kempe, 1993) showed that in the usual hydrological functioning, the relatively clear waters of the Upper Paraná mostly transported a comparatively organic-rich suspended load (4–10% of particulate organic carbon, POC), with noticeably labile organic matter (20–40% of POC), dominantly composed of phytoplankton (e.g. mannose:xylose ratios of 1.8–2.6), with lesser contributions from grasses and zooplankton (e.g. arabinose + galactose relative abundances ≅ 20–25%, calculated on a glucose-free basis). In contrast, the Paraguay River, the system's most significant sediment and POC supplier, delivered a more refractory load (~18% of POC) with higher-than-Upper Paraná's contributions from bacteria and soil-derived materials (e.g. hexose : pentose ratios ≅ 1.5). Finally, the middle reach and the flood plain appear to contribute labile POC and also material that has undergone intense biological degradation, as suggested by the relative abundances of ribose (~21%) and arabinose + galactose (~19%). The mostly autochthonous nature of Paraná's POC is also indicated by the $\delta^{13}C$ of POC, which has produced values in the range –20 to –25‰ in a limited number of determinations.

DISSOLVED COMPONENTS

In this contribution the data gathered during the development of a Scientific Committee on Problems of the Environment (SCOPE) and United Nations Environment Programme (UNEP) joint project is reviewed, which aided in the collection of samples in the Paraná River between March 1981 and November 1984. During this period, the 1982–1983 ENSO event triggered the highest flood ever recorded in the Paraná. The analytical methodology and sampling procedures have been reported elsewhere (Depetris & Kempe, 1990, 1993).

The first assessment of the data (Depetris & Kempe, 1990, 1993) showed that dissolved organic carbon (DOC) was the only major biogeochemical variable whose mass transport rate increased significantly (2.6 times) during the ENSO flood.

Figure 2 shows the variability of DOC in the middle Paraná River. On this occasion, data was separated into two sets: the concentrations measured during the rising and descending limbs of the hydrograph. The highest DOC concentrations were most likely supplied by the river's upper reaches (i.e. the tropical Gran Pantanal and Brazil's Mato Grosso) in an initial phase of the flood. DOC measured when discharges were decreasing and hence, water was draining from the inundated valley, mainly represents the contribution of the dissolved organic species stored in ponds, abandoned channels, etc. in the islands and flood plain. Lower concentrations associated with high discharges were linked to final, more diluted stages of the exceptional flood.

In Fig. 3(a) the procedure was repeated but with the CO_2 partial pressure (PCO_2), which is indicative of bacterial respiration. In contrast with Fig. 2 it shows that the highest PCO_2 concentrations were determined when inundated areas were drained and CO_2-supersaturated water was expelled from the riparian system and injected into the river's main stem. PCO_2-depleted waters were associated with the highest discharges.

Figure 3(b) presents the variability of dissolved inorganic carbon (DIC) during the sampling period. Since total CO_2 or DIC = $CO_{2\,(free)} + HCO_3^- + CO_3^{2-}$, differences between Fig. 3(a) and (b) can be solely attributed to the alkalinity contribution because the carbonate species concentration is always close to zero. Again, highest DIC concentrations were measured during the descending limb of the hydrograph, when water was leaving the flooded islands and river valley. It is noteworthy that the highest DIC concentrations were measured al lower discharges than the highest PCO_2 concentrations. This is due to a more important relative contribution of (organic?) alkalinity to the DIC composition in the initial stages of the flushing out of lentic environments.

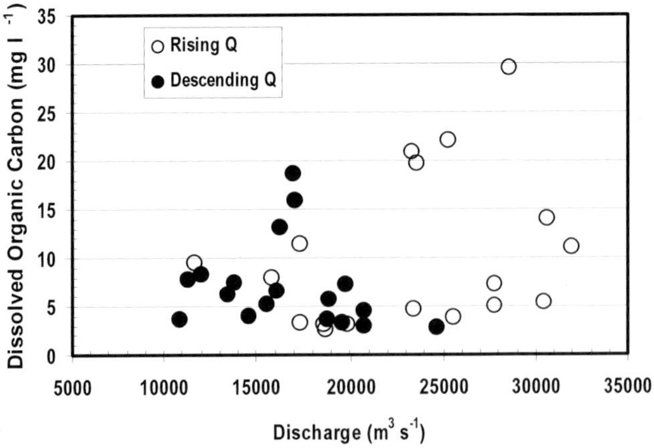

Fig. 2 DOC and discharge relationship in the Middle Paraná River at Paraná, about 600 km upstream from the mouth. Mean annual discharge is ~16 000 $m^3\,s^{-1}$.

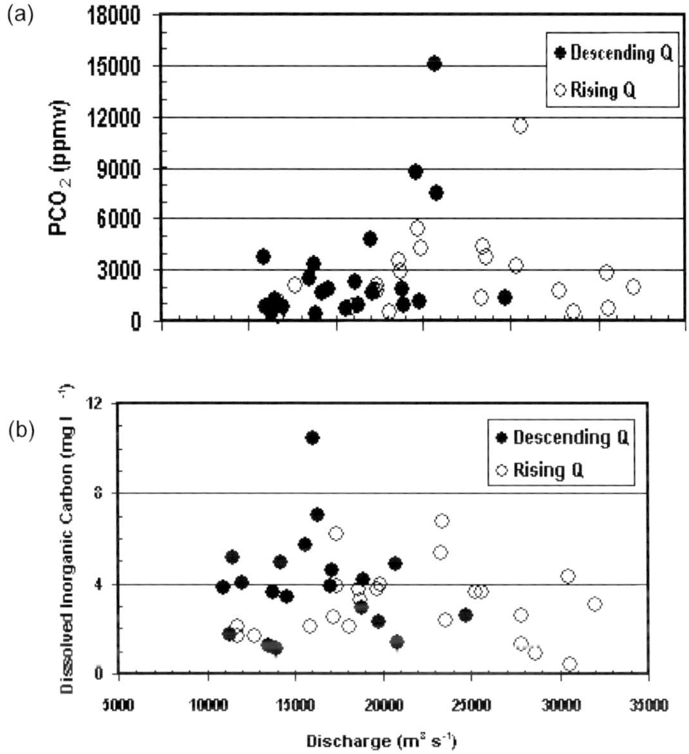

Fig. 3 (a) Relationship between CO_2 partial pressure and discharge; (b) Relationship between DIC (or total CO_2) and discharge. Samples collected at Paraná, about 600 km upstream from the mouth. Mean annual discharge is ~16 000 $m^3 s^{-1}$.

PARTICULATE PHASES

Figure 4 shows the relationship, observed during the study period, between total suspended solids (TSS) concentrations and discharge in the Middle Paraná River. High TSS concentrations were measured at higher-than-the-mean discharges (~16 000 $m^3 s^{-1}$) but also when flood water abandoned the inundated flood plain. Hence, the transported sedimentary material had different origins: the former was probably largely supplied by the headwaters; the latter was the added result of re-suspension of detrital material initially accumulated in the flood plain plus the TSS load from upstream. Further, Fig. 4 denotes that during the high flood a process of dilution occurred when discharge increased markedly. However, the TSS concentration decrease at very high discharge values was also determined by a several-fold increase of the channel width (which resulted in a water velocity decrease), by the trapping effect of the submerged vegetation, and also, by a marked decrease in the surface water slope (e.g. down to ~1 cm km^{-1}, Depetris & Gaiero, 1998) that affected the water velocity and the sediment carrying capacity. All these factors determined the accumulation of sediment in the river's lower reaches during over bank flow, which was probably largely re-suspended when the flood began to subside, and the water surface slope increased again (e.g. to ~2.5 cm km^{-1},

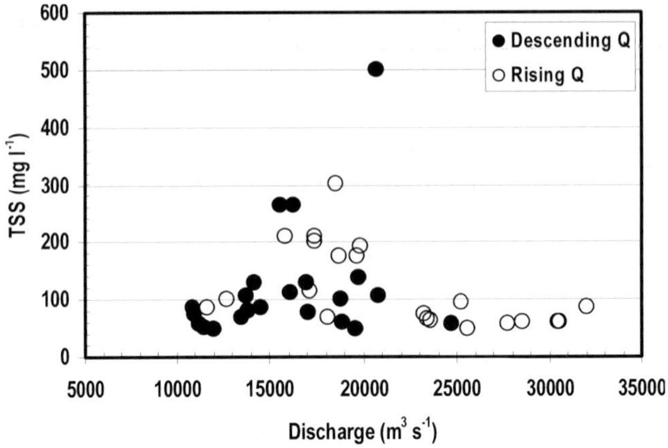

Fig. 4 Total suspended solids (TSS) concentration and discharge relationship at Paraná, about 600 km upstream from the mouth. Mean annual discharge is ~16 000 m^3 s^{-1}. Paraná's mean sediment load is ~80 Tg $year^{-1}$, which corresponds to a mean TSS concentration of ~160 mg l^{-1}.

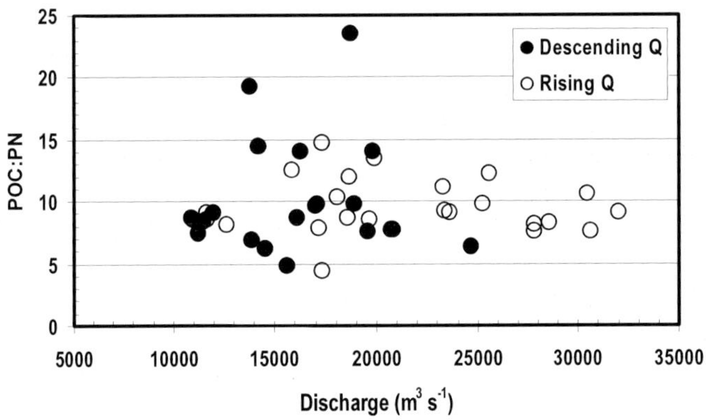

Fig. 5 Carbon to nitrogen ratio (POC:PN) in TSS, determined in the Paraná River, about 600 km upstream from the mouth.

Depetris & Gaiero, 1998) along with water velocity. All in all, there appeared to be little difference between the sediment mass transported before the flood and during its occurrence (Depetris & Kempe, 1990).

The widely accepted Paraná's mean sediment transport rate is ~80 Tg $year^{-1}$ (e.g. Milliman & Syvitski, 1992). Long-term measurements indicate that about 5% (i.e. 4 Tg $year^{-1}$) of such annual load is particulate organic carbon (POC), which is dominantly supplied to Paraná's lower reaches by the Paraguay River. During the 1982–1983 ENSO extreme flooding, the DOC transport rate increased only slightly to 8 Tg $year^{-1}$. It is most likely then that a sizeable proportion of the increase (~4 Tg $year^{-1}$) is attributable to the organic material that was stored and biologically-processed in the flood plain during the inter-flooding period.

The aspect addressed above is illustrated in Fig. 5, which shows the variation of the carbon to nitrogen ratio (POC : PN) in Paraná's suspended load. Highest values occurred when the river discharge was higher than the long-term mean, i.e. within the 15 000–20 000 $m^3 s^{-1}$ range. Such ratios (~15–25) suggest that, under such hydrological conditions, the water that drained the insular system transported a considerable proportion of fresh and labile vascular plant debris (i.e. plant leaves, branches, floating vegetation). High discharges occurring during the flood are linked to POC : PN ratios ≈10, which is normal for soil humic material and planktonic tissue (POC : PN <8).

CONCLUDING COMMENTS

In this contribution I have revisited data collected in the Paraná River several years ago (Depetris & Kempe, 1990, 1993), with an emphasis placed on examining the role played by the river stage on selected biogeochemical variables. In particular we examined their response when the flood water was entering the flood plain, overflowing the numerous lentic and lotic water bodies, and when the flood was in a receding stage and the valley was draining. DOC, being the only biogeochemical variable whose mass transport rate increased significantly during the 1982–1983 ENSO flood, also appears to be the only one that exhibited high concentrations (≥20 mg l^{-1}) at very high discharges. Such high DOC values appeared during the rising limb of the hydrograph, at the initial stages of the flood and were, most likely, supplied by the Paraná's tropical headwaters. The lower DOC concentrations, also measured during very high discharges, were supplied during the final stages of the flood.

The remaining variables, PCO_2 and DIC, reached highest concentrations during flood recession and appeared to be affected by the flood plain and the water bodies therein. Similarly, the POC : PN ratio in the TSS reached the highest values during the over bank stage (15 000–20 000 $m^3 s^{-1}$), thus suggesting that freshly produced vascular plant material was being introduced into the river's main stem.

These mechanisms operating during very high floods in large river systems must be fully recognized if we wish to interpret the changes occurring in the rivers themselves and also the impact generated on the receiving estuaries during such extraordinary events.

Acknowledgements I would like to thank A. I. Pasquini for her assistance in the drafting of this paper's figures. I am also grateful to Argentina's CONICET and FONCYT for sustained support. This contribution was financed by FONCYT's PICT 07-08524.

REFERENCES

Amarasekera, K. N., Lee, R. F., Willams, E. R., Eltahir, E. A. B. (1997) ENSO and the natural variability of flow of tropical rivers. *J. Hydrol.* **200**, 24–39.

Committee on Flux of Organic Carbon to the Ocean (1981) *Carbon Dioxide Effects Research and Assessment Program. Flux of Organic Carbon by Rivers to the Oceans.*CONF-8009140, US Department of Energy, Washington, DC, USA.

Depetris, P. J. & Gaiero, D. M. (1998) Water-surface slope, total suspended sediment and particulate organic carbon variability in the Paraná River during extreme flooding. *Naturwissen.* **85**, 26–28.

Depetris, P. J. & Kempe, S. (1990) The impact of the El Niño 1982 event on the Paraná River, its discharge and carbon transport. *Paleogeogr., Palaeoclim., Palaeoecol. (Glob. Planet. Chan. Sect.)* **89**, 239–244.

Depetris, P. J. & Kempe, S. (1993) Carbon dynamics and sources in the Paraná River. *Limnol. Oceanogr.* **38**, 382–395.

Depetris, P. J., Kempe, S., Latif, M. & Mook, W. G. (1996) ENSO-controlled flooding in the Paraná river (1904–1991). *Naturwissen.* **83**, 127–129.

Degens, E. T., Kempe, S. & Richey, J. E. (1991) *Biogeochemistry of Major World Rivers.* John Wiley & Sons, Chichester, UK.

Genta, J. L., Pérez-Iribarren, G. & Mechoso, C. R. (1998) A recent increasing trend in the streamflow of rivers in southeastern South America. *J. Climate* **11**, 2858–2862.

Kousky, V. E., Kagano, M. T. & Cavalcanti, I. F. A. (1984) A review of the Southern Oscillation: oceanic–atmospheric ciculation changes and related rainfall anomalies. *Tellus* **36A**, 490–504.

Millliman, J. D. & Syvitski, J. P. M. (1992) Geomorphic/tectonic control of sediment discharge to the oceans: the importance of small mountainous rivers. *J. Geol.* **100**, 525–544.

Richey, J. E., Hedges, J. I., Devol, A. H., Quay, P. D., Victoria, R., Martinelli, L. & Forsberg, B. R. (1990) Biogeochemistry of carbon in the Amazon River. *Limnol. Oceanogr.* **35**, 352–371.

Robertson, A. W. & Mechoso, C. R. (1998) Interannual and decadal cycles in river flows of southeastern South America. *J. Climate* **11**, 2570–2581.

Robertson, A. W. & Mechoso, C. R. (2000) Interannual and interdecadal variability of the South atlantic convergence Zone. *Mon. Weath. Rev.* **128**, 2947–2957.

Ropelewski, C. F. & Halpert, M. S. (1987) Global and regional scale precipitation patterns associated with the El Niño/Southern Oscillation. *Mon. Weath. Rev.* **115**, 1606–1626.

The characteristics of geochemistry in a headwater wetland, Chiba, Japan

CHANGYUAN TANG[1] & YASUO SAKURA[2]

1 *Graduate School of Science and Technology, Chiba University, 1–33 Yayoi, Inage, Chiba 263-8522, Japan*
cytang@faculty.chiba-u.jp

2 *Earth Science Department, Faculty of Science, Chiba University, 1–33 Yayoi, Inage, Chiba 263-8522, Japan*

Abstract From the hydrochemical viewpoint, a wetland functions both as an interface zone for water exchange between groundwater and surface water, and as a biochemical filter through which water quality changes greatly. In order to understand clearly how and where water quality evolution processes happen in headwater wetlands, a typical headwater in Chiba, Japan, was chosen to investigate nitrate behaviour in groundwater. Major ions, DO, Eh and $\delta^{15}N$ of groundwater from six groups of piezometers (downslope transect, L1–L5), and a spring (L6) were measured. It was found that Ca^{2+}, Mg^{2+} HCO_3^- and NO_3^- concentrations decreased from L1 to L5 and increased from L5 to L6. SO_4^{2-} had an opposite behaviour to NO_3^-The pe-pH diagram was also used to explain the reduction status in the wetland. As a whole, the water flow system controlled the chemical evolution procedure in the wetland, where the reduction sequence changed from nitrate (from L1 to L4) to sulfate (from L4 to L5) when they migrated in the headwater wetland.

Key words denitrification; groundwater; Chiba, Japan; nitrate; wetland in headwater

INTRODUCTION

From the viewpoint of ecohydrology, a wetland at the headwater is the symbol of the discharge area for the groundwater water system and supports habitats of high bio-diversity. As a result, groundwater is important not only for the formation of wetlands but also as a major water source for rice production in Asia. Wetland is an important part of the headwater zone and is the discharge area linking groundwater and surface water in the basins of humid regions such as Japan. Also, it functions both as an interface zone for water exchange between groundwater and surface water, and as a biochemical filter through which water quality changes greatly.

Nitrate, the principal nitrogen-bearing constituent of groundwater, is found mainly in agricultural regions as a result of the widespread application of synthetic fertilizers and animal manure to agricultural land. Excess nitrogen promotes NO_3^- leaching to groundwater that, when entering surface water, affects the aquatic environment, particularly through eutrophication of coastal waters. Moreover, NO_3^- is a danger to human health, especially for pregnant women and children, and so poses a serious threat to drinking water supplies. In view of these hazards, specific land-use manage-ment is needed. One approach consists of promoting effective methods for *in situ* removal by denitrification, i.e. the transformation of NO_3^- into gaseous N_2 or N_2O. Nitrate removal in wetlands has been attributed to this process, and this is the reason why the management of natural as well as constructed wetlands is receiving increasing attention. For a long time it has been well known that wetlands and paddy fields have some functions in the decline of nitrate content in water. However, this knowledge only concerns the biological processes such as nitrate uptake by plants, or the anaerobic conditions developed in water pools usually found in wetland and paddy fields. In order to understand how and where water quality evolution processes happen in the headwater wetland, a typical headwater in Chiba, Japan, was chosen to investigate nitrate behaviour in groundwater. This paper focuses attention on:

(a) identifying the redox conditions that control the behaviour of nitrate at the headwater wetland;
(b) study of the changes of chemicals in groundwater when it moves downstream through the wetland; and
(c) comprehensive understanding of relation between groundwater flow and the changes of its chemical components.

HYDROLOGICAL SETTINGS

The study area is located at Ochi-cho, Chiba City, Japan. Geomorphologically, the study area is composed of the upland, the slope and the valley (Fig. 1(a)). Vegetables are planted in the former, and the latter contains paddy fields for rice production and is also an important residential area in Chiba. The upland is the recharge area in the study area where there are farmlands for vegetables. People living there use groundwater for drinking (W1, W2 and W3). The slope is covered by trees and the rice paddy fields can be found at the lowland of the valley. The altitude difference between the upland and lowland is about 25 m.

Groundwater piezometers were installed in a transect, A–B, from the upland margin through the headwater wetland to the spring (Fig. 1(b)). From the viewpoint of hydro-geomorphology, the wetland can be divided into three zones (listed upstream to downstream), the shallow water table zone (between L1 and L3), the seepage zone (between L3 and L6) and the spring zone (at L6). Six piezometer groups were set along the A–B cross-section. There is seepage water in the zone between L4 and L6, and a spring at L6 all year round. Seepage can be found seasonally in the area between L3 and L4 from summer through autumn when the water table is high in the wetland.

From the surface to 1.0–1.2 m depth there is a loam layer underlain by a silty sand layer. At L6, the loam layer disappears and only the silty sand layer exists from which groundwater shows up as the spring that supplies water for crop production at nearby paddy fields on the downstream side.

Field surveys were conducted once or twice a month from May 2000 to February 2002. The hydraulic potentials were measured at piezometers. Groundwater samples were taken at the piezometer locations and spring water was sampled at point L6 once a month. In order to monitor seasonal variations of nitrate concentration of wells in the upland, samples were taken once every three months. During field surveying, pH, electrical conductivity, temperature, DO and redox potential of the groundwater were measured *in situ*. Water samples were analysed in the laboratory at Chiba University. Anions such as NO_3^-, Cl^- and SO_4^{2-} were measured using ion chromatography (LC-10A,

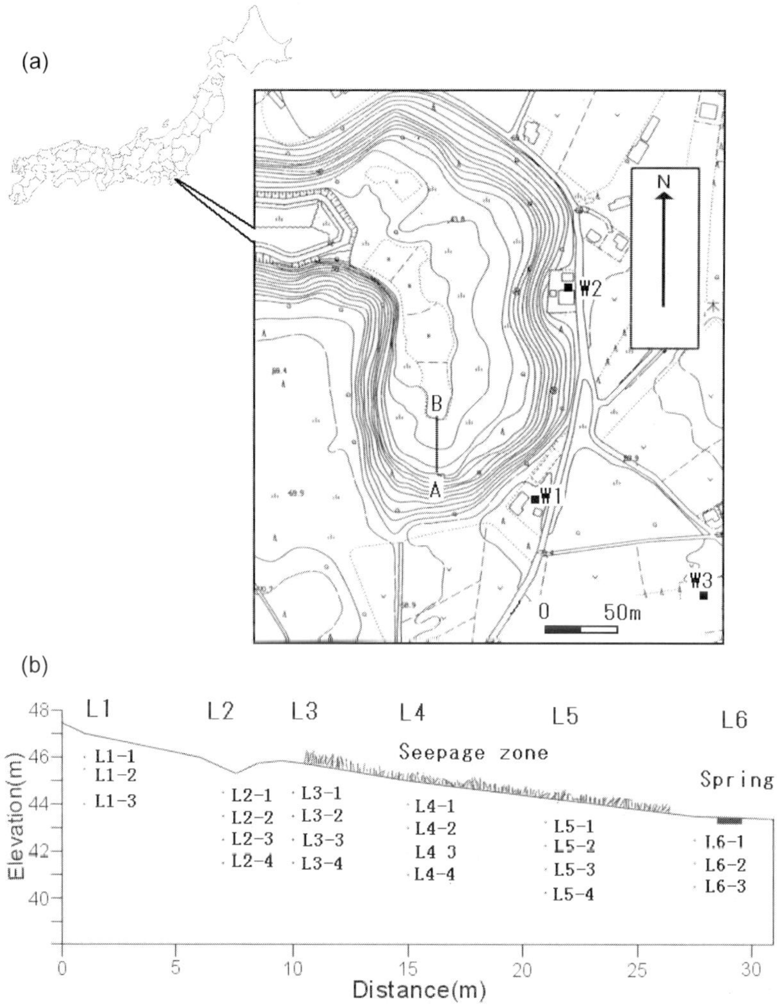

Fig. 1 (a) Map of the study area. (b) Piezometers in the cross section A–B.

Shimadzu) and cations such as Na^+, K^+, Ca^{2+} and Mg^{2+} by standard inductively coupled plasma atomic emissions spectrometry (ICPS-1000 III C, Shimadzu). The concentration of HCO_3^- was measured using titration.

NO_3^- bearing water samples were prepared for ^{15}N isotope measurements by the standard procedure, i.e. transformation of NO_3 to $(NH_4)_2SO_4$, the speciation of N_2 from the reaction of H_3PO_4 with $(NH_4)_2SO_4$, and the purification of N_2 gas through a vacuum glass line. ^{15}N isotope was analysed by the mass spectrometer (Delta S, Thermoquest). $\delta^{15}N$ values are expressed as the per mill (‰) difference of the isotope ratios of a sample (sp) and a standard (std), referred to as AIR for ^{15}N. They are defined as follows:

$$\delta^{15}N = \left(\frac{R_{sp} - R_{std}}{R_{std}}\right) \times 10^3 \quad ‰ \qquad (1)$$

where R is the ratio of the heavy to the light isotope, that is, $R = (^{15}N/^{14}N)$. Average deviations from the mean $\delta^{15}N$ values for samples were in the order of ±0.1–0.3‰.

RESULTS AND DISCUSSION

Characteristics of nitrate distribution in groundwater

Several recent studies have found that larger groundwater nitrate reservoirs exist in catchments than previously suspected (Kendall *et al.*, 1995a,b; Williams *et. al.*, 1997; Burns *et al.*, 1998). We have taken some samples from the wells in the upland and measured the nitrate concentrations. It was found that the groundwater in W1, W2 and W3 were high in nitrate (from 38–58 mg l^{-1}) and small in $\delta^{15}N$ values (+3.8–4.2‰). Groundwater flowing into the wetland also had the same characteristics with high nitrate and small $\delta^{15}N$ values (L6 (3 m), L1 (1 m)). It can be explained that, in general, nitrate in groundwater mainly came from the fertilizers used in the upland, the recharge area of the study site.

Figure 2 shows the distribution of average NO_3^- concentrations along the cross section A–B at the study site. Before groundwater moved into the wetland, NO_3^- concentrations in the groundwater were high. When groundwater moved across the headwater in the study site, it was found that there were three types of vertical profiles of nitrate in groundwater, i.e. concentration decreasing with depth (L1 and L2), increasing with depth (L6), and low concentrations at all depths (L3 to L5). Actually, the first type was located at a zone where groundwater flowed downward or horizontally, while the second one was found in the zone with upward groundwater flow. From L2 to L3, where the groundwater flowed horizontally, nitrate concentrations in the groundwater decreased dramatically to 7/10 at 3 m depth and to an undetectable level at 4 m depth. The decline of nitrate concentration in groundwater was also found in the piezometers at 1 and 2 m depth. This indicates that nitrate removal occurred in a very narrow zone before groundwater had moved into the seepage zone. Nitrate was not detected at all the piezometers at L5. When groundwater moved upward at L6, nitrate concentrations decreased a little between 3 and 2 m depth, but by almost 1/3 between 2 and 1 m depth.

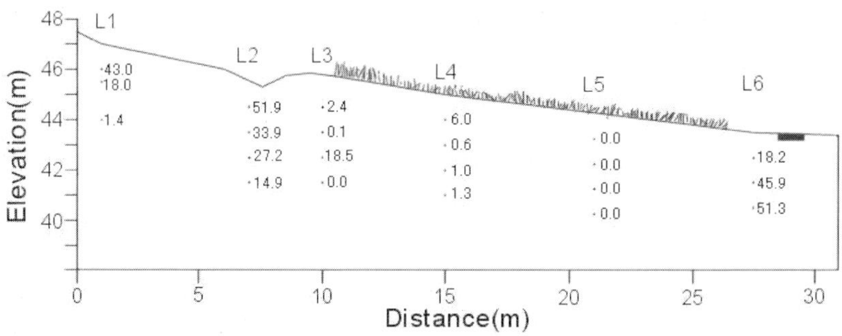

Fig. 2 Distributions of nitrate concentration (mg l^{-1}) in piezometers along the cross section A–B.

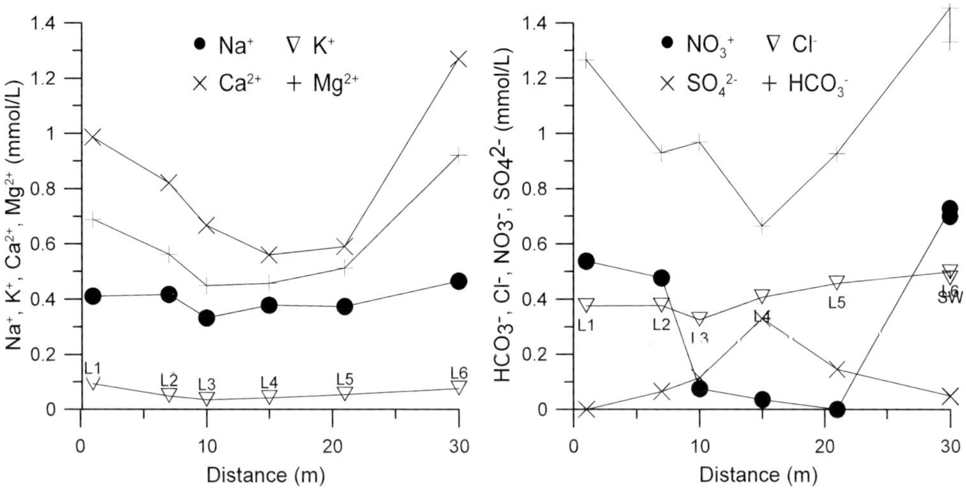

Fig. 3 Variations of average concentrations of major ions along the flow path.

The variations of major ions in the groundwater

Figure 3 shows the variations of average concentrations of anions and cations from L1 to L6 at the study site. The variations of K^+, Na^+ and Cl^- concentrations were small along the flow path of the wetland. However, Ca^{2+}, Mg^{2+} HCO_3^- and NO_3^- concentrations decreased from L1 to L5 and increased from L5 to L6. The SO_4^{2-} concentration had an opposite behaviour to NO_3^-. The SO_4^{2-} concentration was low before the groundwater flow into the headwater. From L1 to L4, the negative linear relation between nitrate and sulfate may be most conveniently explained by denitrification caused by the oxidation of sulfides:

$$14NO_3^- + 5FeS_2 + 4H^+ \rightarrow 7N_2 + 10SO_4^{2-} + 5Fe_2^+ + 2H_2O \quad (2)$$

From Fig. 3 we can calculate the concentration changes of nitrate relative to the changes of sulfate between L1 and L4 using the regression equation:

$$[SO_4^{2-}] = -1.5024[NO_3^-] + 0.4725 \qquad (r = 0.82) \quad (3)$$

Using equation (3), the ratio equals –1.6 mmol NO_3^- per mol SO_4^{2-}. According to the reaction equation (1) this ratio equals –1.5. This means that the greater part of nitrate is apparently denitrified as shown in equation (2), the remaining part, 0.1 mmol NO_3^-/mmol SO_4^{2-} being denitrified by other processes.

Sulfate reduction was also present between L4 and L5. The sulfate reduction is due to the influence of nitrate in the groundwater. When the groundwater was almost free of nitrate at L4, sulfate became the new electron accepter in the system. Therefore, reduction conditions developed in the wetland with the reduction sequence for nitrate (from L1 to L4) and sulfate (from L4 to L5).

However, sulfate concentrations became low at point L6 where groundwater flowed up as a spring with high nitrate concentration. As a result, the spatial distribution of major ions show that the wetland can be divided into two parts, one from L1 to L5 and the other at L6, because of the two water flow paths in the study area.

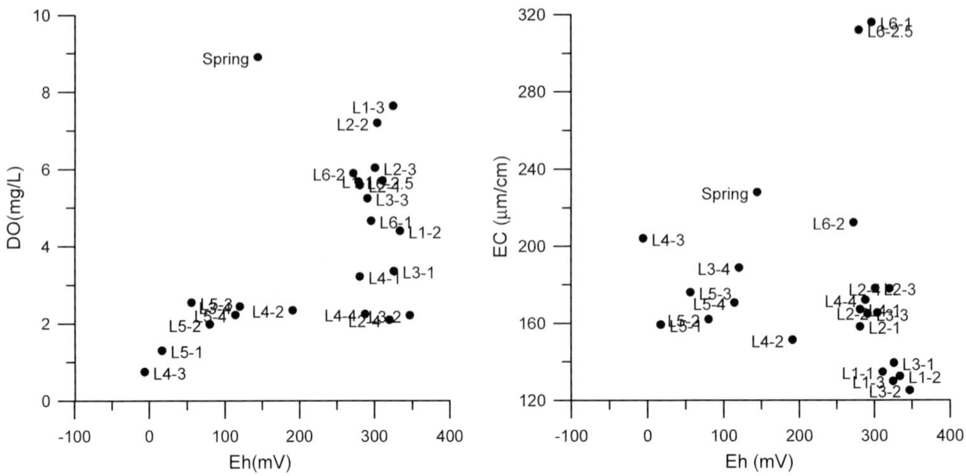

Fig. 4 Relationships between Eh and DO (left), Eh and EC (right) in the study area.

Behaviour of DO, EC, pH and Eh in the wetland

Figure 4 shows the relationships between Eh and DO, as well as that between Eh and EC, in the study area. When the groundwater flowed from L1 to L3, DO values decreased sharply ranging from 7.6 mg l^{-1} at L1–L3 to 2.1 mg l^{-1} at L2–L4. At the same time, the variations of Eh values ranged from 280 to 350 mV. From L3 to L4, DO values stayed around 2–3 mg l^{-1} with a sharp decline of Eh. From L4 to L5 both Eh and DO decreased along the groundwater flow path. However, groundwater and spring water at L6 were high in DO.

The EC values of the spring water and groundwater at L6 were larger than that of groundwater from L1 to L5 (Fig. 4). The explanation is that there was an upward groundwater flux at L6. Nitrate derived from fertilization of the upland area moved down to the headwater wetland at concentrations of over 50 mg l^{-1}. Since the oxygen levels were fairly low in the wetland, the oxygen was rapidly depleted and, as the level fell to, and then below 2 mg l^{-1}, nitrate was used as an electron acceptor. As a result, nitrate is reduced through the unstable intermediate product nitrite, to N_2 in the process known as denitrification.

Relationships between species containing oxidized and reduced members of redox couples are commonly displayed on diagrams know as pe-pH diagrams. The widespread occurrence of aquifers contaminated by nitrate, particularly in agricultural areas, requires a close look at redox transformations of nitrogen in groundwater. Nitrogen, an essential nutrient for crop growth, also occurs in human and animal waste and is the most abundant gas in the atmosphere. Nitrogen occurs in various oxidation states from nitrate to ammonia. Figure 5 shows stability fields for aqueous nitrogen species as functions of pe and pH. The measurement data from the site were also plotted in the diagram. It was found that all points in the study area fell into the part where denitrification happened. If we look more closely at their distribution (enlarged diagram top right Fig. 5), the pe decreased from L1 to L5 along the flow path in the study area. It is notable that, nitrate reduced rapidly from L2 to L4 with decrease of pH

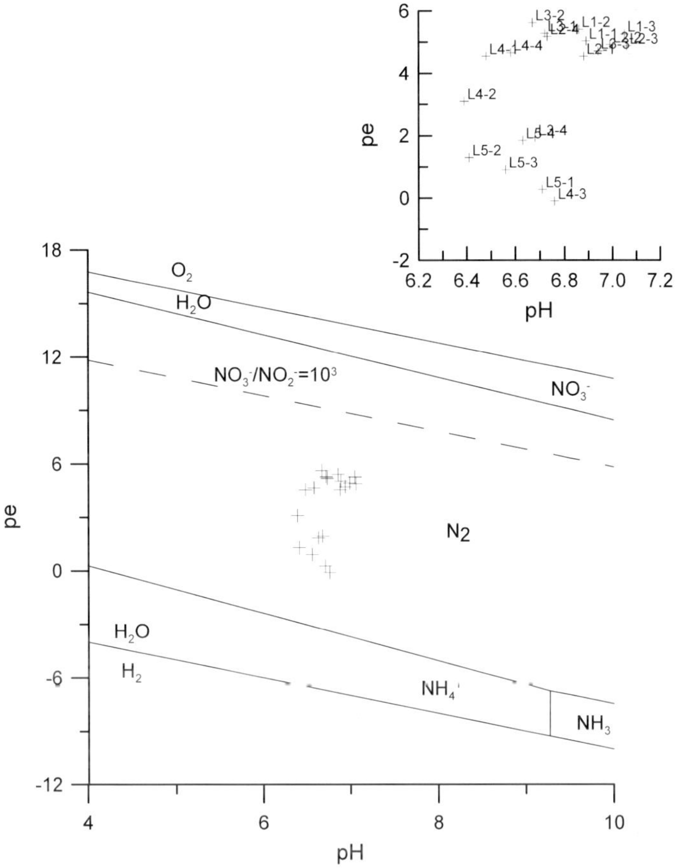

Fig. 5 pe-pH diagram for selected species in the N-O_2-H_2O system at 25°C; data from the study area are plotted in the diagram.

value, which also was supported by the chemical reaction equation (1). Considering the data shown in Figs 3, 4 and 5, when nitrate travelled through the headwater wetland, the redox potential dropped along with the removal of dissolved oxygen in groundwater. As a result, nitrate functioned as an electron acceptor and was reduced to N_2 in the process of denitrification. At the same time, transformations between species were almost exclusively facilitated by micro-organisms. Thus, the equilibrium relationships implied by a pe-pH diagram must be understood to be controlled by microbial kinetic factors.

Effects of groundwater flow on the distribution of chemical components in the wetland

The groundwater flow system is the key to understanding the migration of chemicals in the wetland system. The groundwater in the study site was recharged at the upland and moved down the valleyside and flowed up to the surface at the wetland of the headwater. When groundwater flowed downstream from L3, the water table was high enough to make a seepage zone between L4 and L6. Also, the total potential increased with increasing depth between L4 and L6, causing the groundwater to flow toward the surface. It was found that the upward hydraulic gradient between groundwater and the springs was as large as 1.5 m m^{-1} at L6 where there was a spring zone which made a 1.5 m diameter pond that persisted all year round. As a result, there were two paths for groundwater in the study site: the shallow one flowing horizontally across the seepage zone, and another one below an elevation of 42 m moving up towards the spring zone at L6.

Laterally proceeding sequences of redox processes along the flow path are generally found in the wetland of the headwater where groundwater potentials develop a result of discharge zone. When groundwater moves across the wetland, nitrate becomes a very important electron acceptor for bioactivities because of the lower dissolved oxygen content. At the same time, other chemicals such as sulfate and ferric compounds change their concentrations in groundwater, resulting in a net proton production. The change of pH during denitrification indicates an overall proton consumption, which means more acid-base processes are active. As a result, it was found that the chemistry of the water entering the wetland was very different to the chemistry of the water exiting the wetland.

CONCLUSIONS

It was found that there was both shallow groundwater and deep groundwater at the study site. Both groundwater flows joined together and move up to emerge as a spring. Along the flow path between the wells the variations of K^+, Na^+ and Cl^- concentrations were small, Ca^{2+}, Mg^{2+} HCO_3^- and NO_3^- concentrations decreased from L1 to L5 and increased from L5 to L6. However, SO_4^{2-} concentrations decreased as NO_3^- concentrations increased. Nitrate removal occurred in a very narrow zone where DO values were steady at around 2–3 mg l^{-1} with a sharp decline of Eh. Overall, it was found that the DO value has a positive relation with nitrate concentration. Eh, EC and pe data provide strong evidence that reduction conditions developed in the wetland with the reduction sequence for nitrate (from L1 to L4) and sulfate (from L4 to L5) when they migrated in the headwater wetland. The pe-pH diagram for selected species in the system N-O_2-H_2O also shows that the groundwater at the wetland has the strong reducing conditions that make denitrification possible.

Acknowledgements The authors gratefully acknowledge grant-in-aid for scientific researches from the Ministry of Education, Culture, Sports, Science and Technology of Japan, (B)(2) 13480152, and Houga 16651004.

REFERENCES

Burns, D. A., Murdoch, P. S. & Lawrence, G. B. (1998) The effect of ground-water springs on NO_3^- concentrations during summer in Catskill Mountain streams. *Water Resour. Res.* **34**, 1987–1996.

Kendall, C., Burns, D. A., Silva, S. R., Chang, C. C. Y. & McMahon, P. B. (1995a) Sources of variation in the oxygen and nitrogen isotopic compositions of nitrate in soils. *Am. Geophys. Union Trans.* **76**, 210.

Kendall, C., Campbell, D. H., Burns, D. A., Shanley, J. B., Silva, S. R. & Chang, C. C. Y. (1995b) Tracing sources of nitrate in snowmelt runoff using the oxygen and nitrogen isotopic compositions of nitrate. In: *Biogeochemistry of Seasonally Snow-Covered Catchments* (ed. by K. A. Tonnessen, M. W. Williams & M. Tranter), 339–347. IAHS Publ. 228. IAHS Press, Wallingford, UK.

Williams, M. W., Davinroy, T. & Brooks, P. D. (1997) Organic and inorganic nitrogen pools in talus fields and subtalus water, Green Lakes Valley, Colorado Front Range. *Hydrol. Processes* **11**, 1747–1760.

Dynamics and Biogeochemistry of River Corridors and Wetlands (Proceedings of symposium S4 held during the Seventh IAHS Scientific Assembly at Foz do Iguaçu, Brazil, April 2005). IAHS Publ. 294, 2005.

Measuring river health: the uncertainty associated with impact assessment using a single sample biological assessment technique

CATHERINE A. FRANKS[1], MICHAEL J. MAHONY[1] & STEWART W. FRANKS[2]

1 *School of Environmental and Life Sciences, University of Newcastle, Callaghan, NSW 2308, Australia*
catherine.franks@studentmail.newcastle.edu.au

2 *School of Engineering, University of Newcastle, Callaghan, NSW 2308, Australia*

Abstract The AUSRIVAS modelling package is the most sophisticated rapid biological assessment technique available for assessing the health of Australian rivers. Macroinvertebrate samples are collected, and the observed taxa are compared to those expected to be present if the site was in pristine condition. However, the method relies on only a single sample collected from each site to be assessed. Multiple macroinvertebrate samples were collected from sites upstream and downstream of four wastewater treatment plants in the lower Hunter Valley, New South Wales, Australia, to assess the impacts of effluent discharge on creek health, and to investigate the implications of the within-site variability of AUSRIVAS outputs for impact assessment. Preliminary data demonstrate that the AUSRIVAS model can produce a range of outputs for a single site, and that this variability is then amplified creating greater uncertainty when sites are compared to quantify the actual ecological impact of each wastewater treatment plant.

Key words AUSRIVAS models; Australia; effluent discharge; impact assessment; macroinvertebrates; output variability; rapid biological assessment

INTRODUCTION

The growing concern for ecological values in Australia has led to the development of rapid biological assessment techniques for more direct monitoring of water quality and ecosystem health (Norris & Norris, 1995). Macroinvertebrates are small invertebrate organisms (e.g. insect larvae, beetles, molluscs, worms), which are universally abundant in all but the most degraded aquatic habitats (Greenwood *et al.*, 1999). These organisms are particularly suitable as indicators of ecosystem health as macroinvertebrate taxa have a wide range of sensitivities to different types of pollution, and so it is likely that most types of pollution will result in a shift in macroinvertebrate community composition (Chessman, 1995). Macroinvertebrate communities also recover slowly if disturbed due to their sedentary nature and life cycles ranging from a few weeks to a few years (Marchant, 1986), and can retain evidence of pollution incidents long after the disturbance itself has subsided. Finally, macroinvertebrates are small, easy to identify and store, and can be easily collected with simple equipment (Chessman, 1995).

AUSRIVAS (AUStralian RIVers Assessment System) is a predictive modelling package recently developed at the CRC for Freshwater Ecology in Canberra, Australia, as a major component of the National Rivers Health Program (NRHP). Mathematical models are used to predict the macroinvertebrate fauna expected to be present at a site in the absence of environmental stress (Ransom *et al.*, 2003). The models then compare the predicted fauna to the observed fauna to produce a suite of outputs, which can be interpreted to determine the health of the river. However, this paper will focus on only one output, the observed to expected ratio (OE_{50}), as this provides a direct measure of biological impairment. The OE_{50} value is the ratio of the number of macroinvertebrate families observed at a site to the number of families expected (Ransom *et al.*, 2003). The OE_{50} value is then used to classify the site. Upper limits for OE_{50} values are 0.19 for extremely impaired sites (band D); 0.51, severely impaired (band C); 0.83, significantly impaired (band B); and 1.16, reference condition (band A).

It is the aim of this paper to investigate the suitability of AUSRIVAS for monitoring the impacts of wastewater treatment plant effluent discharge on creek health in the lower Hunter Valley, New South Wales, Australia. Since AUSRIVAS derives its outputs using data from just a single macroinvertebrate sample from the site of interest, the within-site variability of OE_{50} values, and subsequent implications for impact assessment, will be examined by comparing multiple samples from within the same sites.

STUDY SITES

There are several inland wastewater treatment plants in the lower Hunter Valley, New South Wales, Australia. The impacts of effluent discharge from four of the larger treatment plants on receiving creek health were investigated. Eight study sites were established in total, one upstream and one downstream of each of treatment plants 1, 2, 3 and 4, which predominantly treat domestic sewage and wastewater, and discharge into Black Creek, Swamp Creek, Anvil Creek and Fishery Creek, respectively (Fig. 1). Each site is defined as a reach length of 100 m, therefore for each treatment plant the total sampling area encompassed 100 m upstream and 100 m downstream of the point of effluent discharge. During periods of drought some of the water from these four creeks is recycled for land irrigation, and so water quality is a particularly important issue, especially since the creeks are relatively degraded due to a long history of mining, agriculture and viticulture in the lower Hunter region. The four creeks are physically similar, exhibiting slow flow velocities, shallow gradients, mean widths of 3–5 m, and large areas of fine silty sediment and deposits of organic matter on the creek beds. Black, Swamp, Anvil and Fishery creeks all ultimately drain into the Hunter River.

METHODS

Field sampling was undertaken during (southern hemisphere) spring 2002, between 15 September and 15 December. Macroinvertebrate samples were collected from each of the eight field sites using standard sampling protocols outlined by Turak & Waddell (2001). However, rather than taking only one sample as the method suggests, ten replicate samples were collected randomly from within each 100-m reach of creek, ensuring that the repeated sampling of any particular area was avoided. No riffle habitats were recorded at any of the sites, and so macroinvertebrates were collected

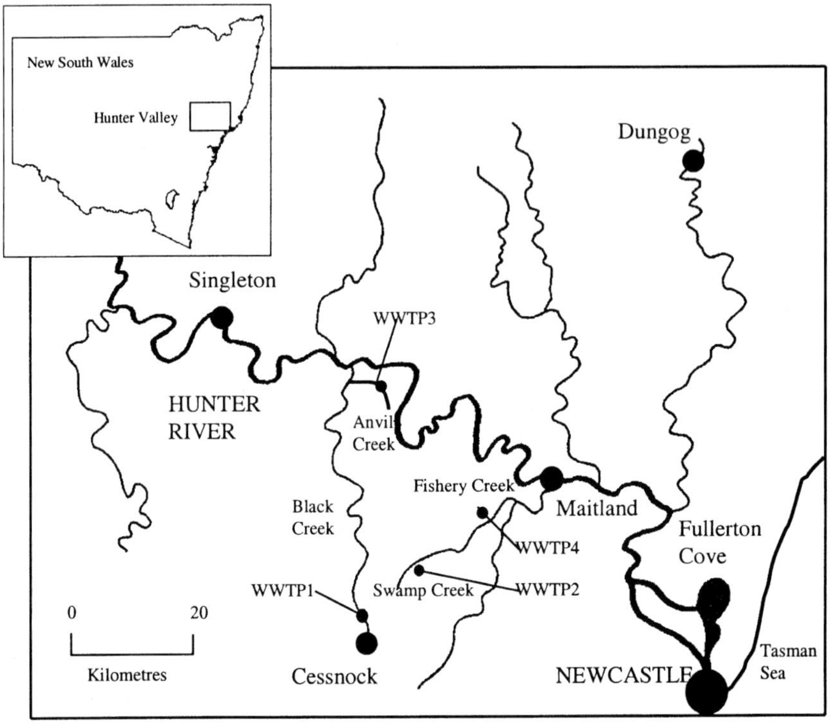

Fig. 1 Map to show the location of wastewater treatment plants 1, 2, 3 and 4 in the lower Hunter Valley, New South Wales, Australia.

only from "edge" habitats, areas along creek banks with little or no current, where leaf-litter, macrophyte beds and trailing bank vegetation may be present (Turak & Waddell, 2001). Collection techniques ensured that surface-dwelling organisms were collected where present, and that the sediment was disturbed so as to suspend benthic organisms for inclusion in the samples. The net was rinsed thoroughly between samples to ensure that no organisms were transferred. Macroinvertebrate samples were live-picked and preserved as soon as possible after collection, following the guidelines outlined by Turak & Waddell (2001). Taxa were identified to the family level where possible using the available keys. At each site a range of geographical variables, channel characteristics and water quality parameters were recorded. Site predictor variables required by the AUSRIVAS NSW spring edge model are alkalinity, altitude, percentage composition of bedrock, boulders and cobbles, latitude, longitude, average rainfall, log distance from source and log slope between the site and one kilometre upstream. These data were then put into the AUSRIVAS NSW spring edge model to produce OE_{50} values for all samples collected. For each wastewater treatment plant, the 10 upstream OE_{50} values were divided by each of the 10 corresponding downstream OE_{50} values to produce 100 u/d OE_{50} ratios (with the exception of treatment plant 4 where only 90 u/d OE_{50} ratios were produced as a result of one downstream OE_{50} value being zero). The distributions of these ratios were then plotted on histograms to demonstrate the potential variability in impact magnitude determination when only single samples from each of the upstream and downstream sites were compared. T-tests were then carried out to identify any significant differences between the means of the 10 OE_{50} values from upstream and downstream of each treatment plant, to determine what, if any, impacts on creek health are actually occurring as a result of effluent discharge.

RESULTS

Table 1 shows the summary statistics for the OE_{50} outputs produced by the NSW spring edge AUSRIVAS model for the macroinvertebrate data collected. These data demonstrate the high variability of OE_{50} outputs that can be produced for samples taken from within the same site. This suggests that even though one macroinvertebrate sample is assumed to be representative of the fauna of the edge habitat in the entire 100 m reach (Turak & Waddell, 2001), the samples are not consistent within-site, implying that one macroinvertebrate sample per site is not enough to accurately assess the status of the water quality and ecological health of these degraded creeks. The variability of within-site OE_{50} values can become amplified causing greater uncertainties in data reliability when one upstream and one downstream OE_{50} value at a particular treatment plant are compared to assess whether any impact on creek health is occurring as a result of effluent discharge. Since 10 macroinvertebrate samples were collected from upstream and downstream of each treatment plant, this gave 100 possible combinations of single-sample upstream-downstream pairs per treatment plant. To quantify the impact implied by each pair, ratios of upstream to downstream (u/d) OE_{50} values were calculated. Figure 2 (a)–(d) show the wide-ranging distributions of these ratios for each treatment plant. A ratio of 1 implies that no discernable impact is occurring, and in this case an arbitrary range of ratios from 0.50 to 1.49 has been defined as the "no impact" range to allow more realistic interpretation of the results. Therefore, u/d OE_{50} ratios greater than 1.49 indicate that some negative impact on creek health is occurring between the upstream and downstream monitoring sites, whereas u/d OE_{50} ratios less than 0.5 indicate some positive impact. As Fig. 2 (a)–(d) demonstrate, u/d OE_{50} ratios from single-sample upstream–downstream pairs for each of the four treatment plants studied span wide ranges, with data for three of the treatment plants representing all impact categories. *T*-tests were conducted on log-transformed data (due to positive skew) to investigate whether any significant differences between mean upstream OE_{50} values and mean downstream OE_{50} values

Table 1 Summary statistics for OE_{50} outputs ($n = 10$).

Treatment plant	Site	Mean	Standard deviation	Range span
1	upstream	0.20	0.09	0.10–0.38
	downstream	0.21	0.11	0.10–0.39
2	upstream	0.43	0.14	0.28–0.71
	downstream	0.21	0.08	0.10–0.29
3	upstream	0.28	0.09	0.16–0.48
	downstream	0.23	0.11	0.16–0.49
4	upstream	0.19	0.17	0.00–0.48
	downstream	0.12	0.08	0.00–0.29

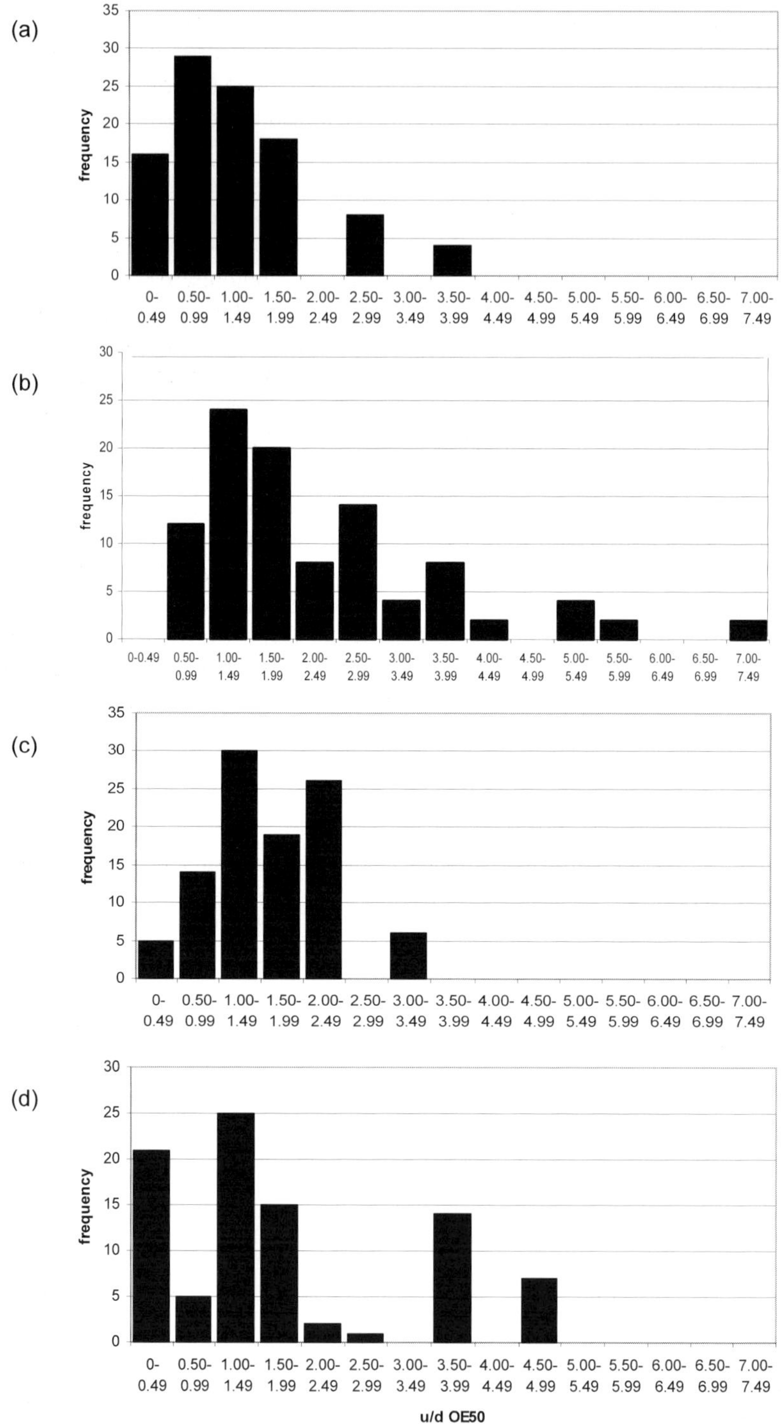

Fig. 2 Histogram of u/d OE_{50} ratios: (a) treatment plant 1, Black Creek; (b) treatment plant 2, Swamp Creek; (c) treatment plant 3, Anvil Creek; (d) treatment plant 4, Fishery Creek.

for each plant were present, to assess whether or not any impacts were actually occurring. Table 2 shows the *P*-values produced by these analyses, and the number of u/d OE_{50} ratios for each treatment plant that fell into the negative, positive and no impact categories. The *P*-values in table 2 demonstrate that only treatment plant 2 exhibited any significant difference in mean OE_{50} values between the ten samples in the upstream and downstream locations ($P = 0.003$, significant at the 0.01 level). Since the mean OE_{50} values for the upstream and downstream sites were 0.43 and 0.21, respectively (Table 1), this suggests that the site upstream of treatment plant 2 exhibits better water quality than the downstream site, and that some negative impact on the health of the Swamp Creek is occurring as a result of the effluent discharge. This is corroborated by the fact that 64% of the u/d OE_{50} ratios produced for treatment plant 2 indicated that a negative impact was occurring. However, the remaining 36% of ratios indicated that no detectable impact on ecological health was occurring in the creek, resulting in a one in three chance that the wrong conclusions could have been inferred about the impacts of treatment plant 2 if only single samples from upstream and downstream were assessed. No impacts of effluent discharge from treatment plants 1, 3 and 4 were apparent in Black Creek, Anvil Creek or Fishery Creek, respectively from *t*-test analyses. However, in each case only 54%, 44% and 30%, respectively, of u/d OE_{50} ratios support this evidence. Therefore, it is highly possible that very different conclusions could have been drawn if the single-sample technique was utilized, and it is suggested that increased numbers of replicate samples or composite samples may provide more robust AUSRIVAS interpretations for comparisons between upstream and downstream sites.

Table 2 Impacts inferred from u/d OE_{50} ratios.

Treatment plant	*P*-value	u/d OE_{50} ratios < 0.5 Positive impact	u/d OE_{50} ratios 0.5–1.49 No impact	u/d OE_{50} ratios > 1.49 Negative impact
1	0.949	16	54	30
2	0.003*	0	36	64
3	0.155	5	44	51
4	0.289	21	30	49

* significant at the 0.01 level.

CONCLUSIONS

Since biological monitoring is becoming increasingly important in river management practice, it is critical that the results of such monitoring can be relied upon so that informed management decisions can be made. The aim of this paper was to demonstrate that in some river systems a single sample of macroinvertebrates is insufficient to adequately represent the entire macroinvertebrate community at a site, and is therefore inadequate to reliably classify the water quality and ecological health of that site. It is suggested that the degraded nature of the creek networks in the lower Hunter Valley results in a more uneven distribution of macroinvertebrates than might otherwise occur in a more pristine environment, and therefore it is harder to collect one single sample that is representative of the entire macroinvertebrate faunal assemblage.

The variability of OE_{50} outputs produced by the NSW spring edge AUSRIVAS models for sites upstream and downstream of the four wastewater treatment plants studied demonstrates AUSRIVAS, a single-sample technique, is inappropriate for reliably assessing the impacts of wastewater effluent discharge in these particular creek networks. Further research will need to be undertaken to establish an optimal number of samples, either replicate or composite, that should be collected from each site to provide a more accurate representation of macroinvertebrate faunal assemblages in these degraded creeks, with the ultimate aim of developing a more suitable method for biologically monitoring the impacts of wastewater effluent discharge on creek health in the lower Hunter Valley.

Acknowledgements Funding from the Hunter Water Corporation and a University of Newcastle Research Scholarship supported this research.

REFERENCES

Chessman, B. C. (1995) Rapid assessment of rivers using macroinvertebrates: a procedure based on habitat-specific sampling, family level identification and a biotic index. *Australian J. Ecol.* **20**, 122–129.

Greenwood, M. T., Bickerton, M. A., Gurnell, A. M. & Petts, G. E. (1999) Channel changes and invertebrate faunas below Nant-Y-Moch dam, River Rheidol, Wales, UK: 35 years on. *Regulated Rivers* **15**, 99–112.

Marchant, R. (1986) Some quantitative aspects of the life history of aquatic insects in temperate Australian rivers. In *Limnology in Australia* (ed. by P. de Decker & W. D. Williams), 151–158. CSIRO, Melbourne, Australia.

Norris, R. H. & Norris, K. R. (1995) The need for biological assessment of water quality: Australian perspective. *Australian J. Ecol.* **20**, 1–6.

Ransom, G., Coysh, J. & Nichols, S. (2003) Macroinvertebrates Bioassessment Predictive Modelling Software V3.0 User Manual. CRC for Freshwater Ecology, Canberra, Australia.

Turak, E. & Waddell, N. (2001) New South Wales Australian River Assessment System Sampling and Processing Manual. EPA, Sydney, Australia.

Water and fluxes of matter transfers during a storm event in the Mengong experimental watershed (Nsimi—South Cameroon)

J. R. NDAM NGOUPAYOU[1], G. R. NKOUE NDONDO[1], J. –L. BOEGLIN[2], J. –J. BRAUN[3], V. B. KAMGANG KABEYENE[1] & G. E. EKODECK[1]

1 *Department of Earth Sciences, Faculty of Sciences, PO Box 812, University of Yaoundé I, Cameroon*
jrndam@uycdc.uninet.cm

2 *LMTG, UMR 5563, 14 Av. Edouard Belin, F-31400 Toulouse, France*

3 *Indian Institute of Science, Department of Metallurgy, Bangalore 560012, India*

Abstract The origin of water and fluxes of matter exported from the small experimental watershed of Mengong during a storm event was determined using a combination of hydrological and hydrochemical methods. For a given storm only 7% of rainfall flows at the outlet of the catchment. The flow rate of the spring does not change during floods and only represents 2% of the streamflow. The spacio-temporal evolution of physico-chemical parameters during a storm event indicates that about 96% of high flow water at the outlet comes from low ground. The chemical composition of water at the outlet is equally dominated by the component from the swampy zone. Low ground marshes play a very active role in the mobilization of dissolved elements and suspended particles through the process of storage and restitution. The principal mechanism of the liberation of water flux from swampy zones is the translatory flow.

Key words equatorial Africa; exfiltration; humid tropical forest; hydrogeochemistry; Mengong catchment; South Cameroon; storage; storm event; swampy area; translatory flow

INTRODUCTION

The different origins of stream water in the catchment greatly influence the hydrology and geochemistry of storm runoff. This behaviour is controlled by the variable inputs of different hydrological, biological, geochemical and physical characteristics. These contributing areas are hydrologically active zones (Dunne & Black, 1970; Mérot, 1988). Their study allows the characterization of the catchment hydrological processes and the assessment of the important role of the saturated areas in particular (Ambroise, 1988; Gascuel-Odoux *et al.*, 1998). The contribution of these different reservoirs and/or contributing areas, can be determined by hydrograph separation of storm flow components using chemical or isotopic tracers (McDonnell *et al.*, 1990; Pionke *et al.*, 1993; Ladouche *et al.*, 1997; Idir *et al.*, 1999) or suspended sediment concentrations (Etchanchu *et al.*, 1986; Probst *et al.*, 1992).

The purpose of this work is: (a) to determine the origin of flows of water and materials in transit at the outlet of the watershed during floods; and (b) to reveal the principal flowing mechanisms which overlap during the genesis of floods in the Mengong catchment. This is possible thanks to a method associating the hydrological and hydrochemical approaches.

The Mengong Small Experimental Watershed (SEW), which represents the ecosystems of the South Cameroon forest, has been retained by the French Ministry of Research as an "ORE/BVET" (Observatoire de Recherche en Environnement / Bassins Versants Experimentaux Tropicaux) since 2003. The objectives of ORE/BVET is the acquisition of various types of data (physical, chemical, biological) and long duration records on the one hand, necessary to follow-up and to model ecological and experimental processes, and on the other hand the implementation in place of long-term experiments which will complete and valorize the tasks of the simple observation of the environment.

MATERIAL AND METHODS OF STUDY

The natural study area

The Mengong SEW whose geographical coordinates are 3°10′N and 11°50′E, is situated in South Cameroon in Nsimi village some 120 km to the SSE of Yaoundé (Fig. 1). With a surface of 0.6 km^2, it belongs to the Nyong River basin (27 800 km^2), which is the most important of the humid tropical forest of the southern Cameroon Plateau. It is made up of two convexo-concave hill slopes between 669 and 703 m high. A swampy zone, which occurs between the two hills, occupies 20% of the basin's surface. It lies on a crystalline substratum, essentially made up of hypersthene granite of the Ntem complex belonging to the northern border of the Congo craton (2900 Ma) (Toteu *et al.*, 1994). A thick pedological cover develops at the expense of the parental charnockite. It is made up of reddish-yellowish ferrallitic soils, whose thickness may be up to 40 m at the summit of the slopes. In the swampy zone, hydro–morphic soils rich in organic material at the top are observed. Climatic influence is of the equatorial type, with four seasons. The inter-annual average of precipitation is 1793 mm. Annual average flow is 383 mm, with an annual average discharge of 7 ± 2 l s^{-1} (calculated from 1993 to 2002). The basin is entirely and permanently covered by a semi-deciduous forest and by farm land. This considerably reduces surface runoff that is confined to the swampy area. The hydrographs of floods usually show a quasi instantaneous response whatever the period of the year, humid or dry. The permanent saturation of the swampy zone explains this rapid response of floods (Ndam Ngoupayou, 1997).

Sampling

The measurement and sampling points during floods are represented in Fig. 1. Sampling was carried out intensively within 24 h (from 8 h 10 min on the 27 April 2002 to 7 h 40 min on the 28 April 2002). The frequency of sampling varied among the sampling points (outlet, bottom swamp and spring). Electrical conductivity (EC), pH, and temperature were measured at the same time as the sampling. Discharge was measured at the outlet and at the spring every 15 min and 30 min, respectively, throughout the flood. A single sample of rainfall (wet depositions) was collected to determine atmospheric contributions.

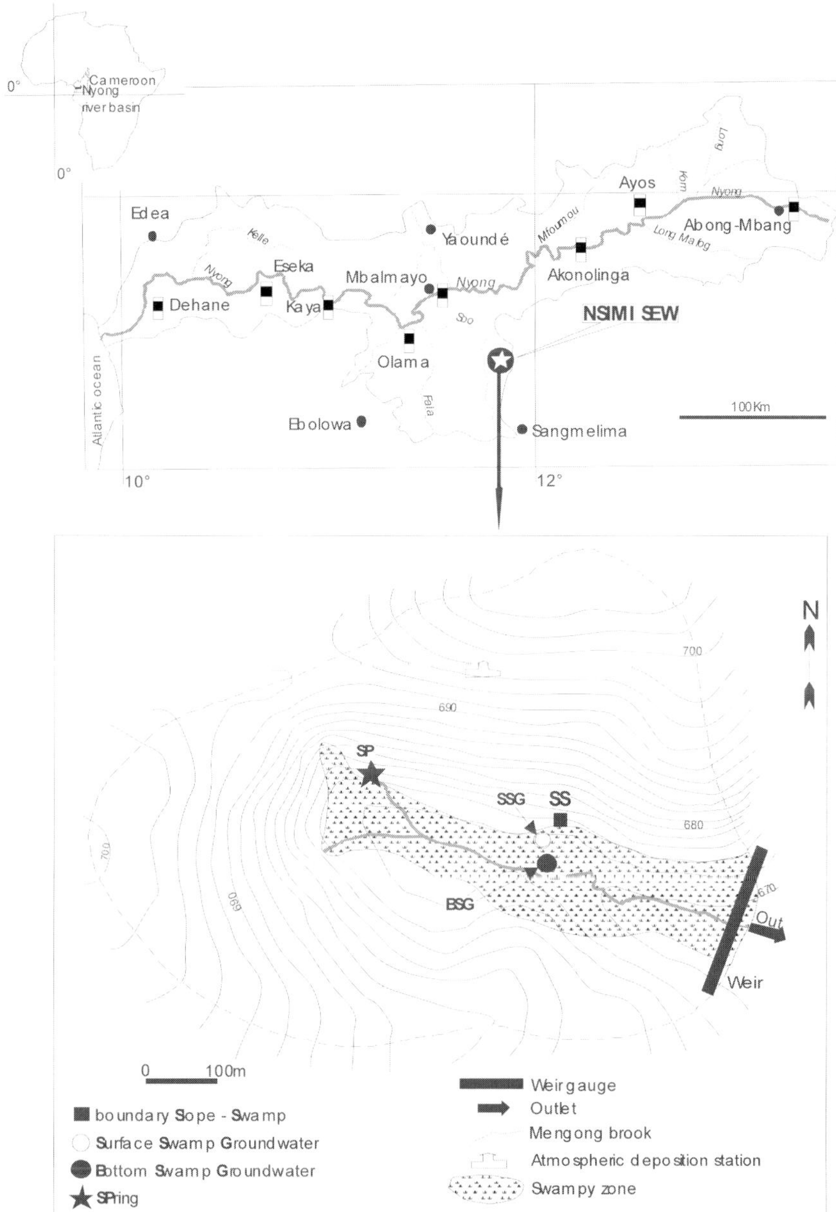

Fig. 1 Location of the Nyong River basin and the Nsimi experimental watershed with the location of the water sampling points [OUT: outlet of the Mengong brook; SP: SPring of the Mengong brook ; SSG: Surface Swamp Groundwater (0.6 m pit); BSG: Bottom Swamp Groundwater (1.5 m pit); SS: boundary Slope-Swamp].

Analytical techniques

The water samples were immediately filtered at the sampling sites through Millipore membranes of 0.45 μm, and conserved in polyethylene bottles at 4°C sheltered from light.

The major anions (Cl^-, NO_3^-, SO_4^{2-}) were determined by liquid ionic chromatography (HPLC-Dionex 4000i) and the major cations (Ca^{2+}, Mg^{2+}, K^+, Na^+) by atomic absorption spectroscopy (Perkin Elmer, Elan 5100). Dissolved silica was measured by colorimetry (Technicon Auto Analyser® II). Dissolved organic carbon (DOC) was measured by catalytical combustion at 680°C using a non-dispersive infrared detector (Shimadzu TOC–5000) and alkalinity was determined by the Gran method. These analyses were all carried out at the Laboratory of Mechanisms of Transfer in Geology (LMTG) and at the Laboratory of Aerology of the Midi Pyrénées Observatory (OMP) at Toulouse, France. The detection limits were 0.3 μg l^{-1} and 8.4 μg l^{-1} for cations and anions, respectively, and 0.1 mg l^{-1} for silica and 0.23 mg l^{-1} for DOC.

RESULTS AND DISCUSSION

Hydrological response of the catchment and origin of flows of water during the flood of 27 April 2002

The flood of 27 April 2002 occurred following a small storm which lasted for about 90 min, 12 mm of rain water on the watershed, thus making the discharge to increase from 1.03 l s^{-1} to 7 l s^{-1}. The response of the basin was very fast. Water started rising less than 10 min after the beginning of the rain and the registered hydrograph presents the characteristics described above.

Discharge at the main spring was extremely low (0.042 l s^{-1}) and did not change in the course of the flood (Fig. 2). This shows that the different functional compartments of the watershed do not react to the storm in the same way. In fact, average discharge registered at the spring in the course of the flood represents only 2% of the total discharge at the outlet. This result leads us to consider an intermediary contributory zone situated between the spring and the outlet, of increasing contribution during the flood. These contributions and the direct precipitation on the channel, can reach up to 96%. The zone thus considered is especially made up of diffused unmeasured contributions, rendering the hydrological approach incapable of explaining the origin of water which flows at the outlet of the watershed during the flood.

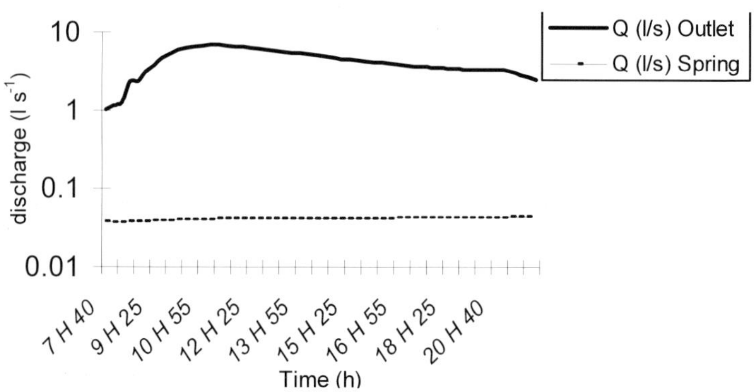

Fig. 2 Hydrographs of floods at the outlet and at the spring of the Mengong catchment in the course of the flood of 27 April 2002.

Evolution of physico-chemical parameters in the course of the flood of 27 April 2002

The measured physico-chemical composition of water in the course of the flood is close to that observed in the earlier studies carried out on the watershed (Ndam Ngoupayou *et al.*, 2002; Viers *et al.*, 1997; Braun *et al.*, 2002; Oliva *et al.*, 1999).

The behaviour of some parameters (pH, DOC, Ca^{2+}, SiO_2 and K^+) has enabled us to graph the mechanisms of flood generation within the watershed.

The pH decreased appreciably at the Outlet (Out) of the basin from the beginning of the flood (Fig. 3). This explains an important contribution of more acidic water.

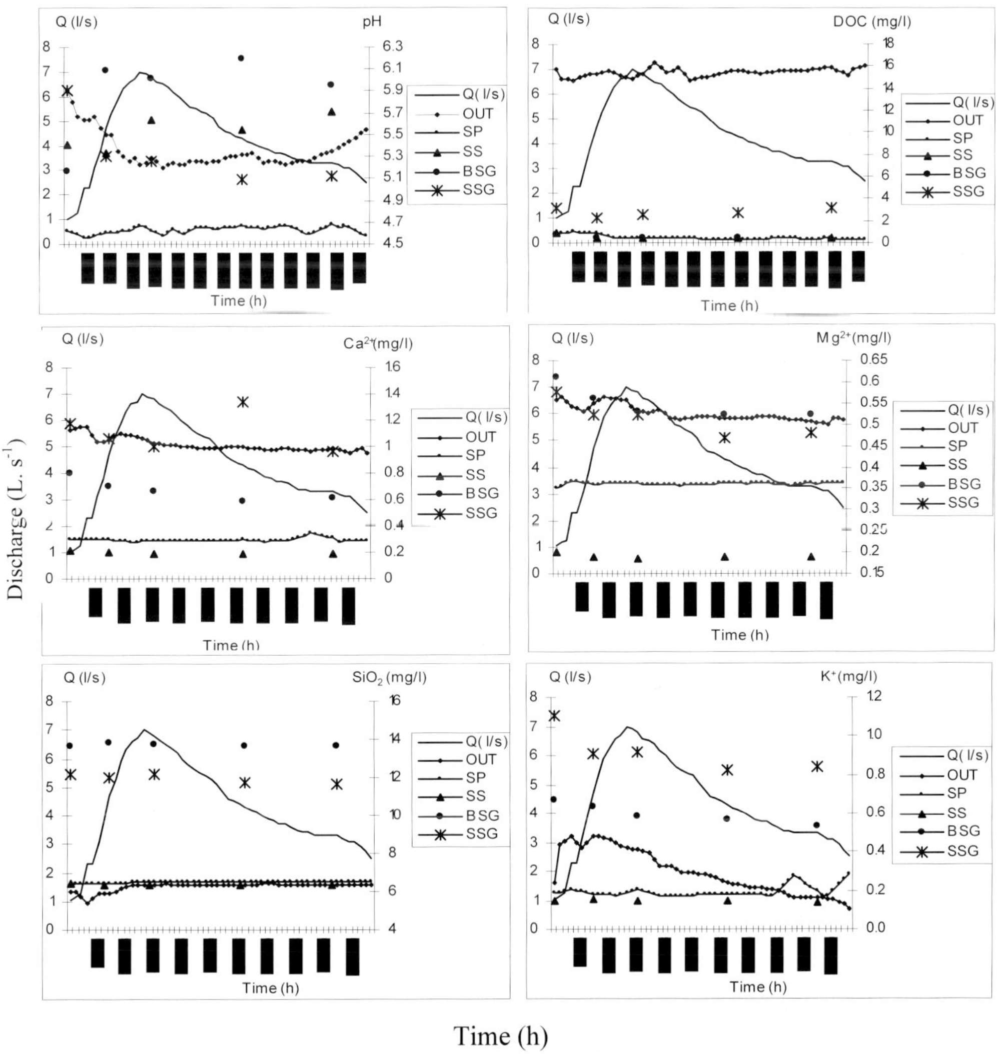

Fig. 3 Evolution of some physico-chemical parameters during the flood of 27 April 2002 [OUT, outlet of the Mengong brook; SP, SPring of the Mengong brook; SSG, Surface Swamp Groundwater (0.6 m pit); BSG, Bottom Swamp Groundwater (1.5 m pit); SS: boundary Slope-Swamp].

This pH is on average 5.5 in the course of the flood and remained very close to the pH of Surface Swamp Groundwaters (SSG = 0.6 m pit) which was on average 5.2. Dissolved Organic Carbon (DOC) at the outlet behaved oppositely to the pH, with an average concentration of 15 mg l^{-1}. This value is similar to that measured at the sub-surface of the swamp not sampled in this study (Ndam Ngoupayou *et al.*, 2002). Variations of Ca^{2+} and at Mg^{2+} are comparable at these two points (outlet and surface swamp groundwater). For these elements, we notice a slight reduction, with almost identical contents (Ca^{2+} = 1.10 mg l^{-1} and Mg^{2+} = 0.53 mg l^{-1} at the outlet, and Ca^{2+} = 1.15 mg l^{-1} and Mg^{+2} = 0.50 mg l^{-1} at the swampy zone). SiO_2 and K^+ contents are higher in both the SSG and BSG ($11.64 < SiO_2 < 13.86$ mg l^{-1} and $0.82 < K^+ < 1.11$ mg l^{-1}) during the flood. Meanwhile, the concentration of these elements at the outlet of the basin are almost identical to those measured at the spring (SP), which are on average SiO_2 = 6.35 mg l^{-1} at the outlet and of 6.45 mg l^{-1} at the spring. This is the same with their temporal variations.

Characterization of contributory zones to the flows of flood: Origin of flows of water and material and their mechanisms of liberation

In the light of observations made on the hydrological and hydrochemical behaviour of water and material fluxes presented in Fig. 3, one can envisage three zones of unequal importance, which are sources of elements at the outlet of the basin, during flooding. The spring, in which the concentration of silica is close to that of the outlet, contributes <2% of the total flow of the flood, thus limiting its influence during the genesis of small floods. The concentration of major elements in rain water is very low compared to the rest of the basin. No real dilutions are observed in the course of the flood. It thus appears that the principal zone responsible for floods is the intermediary zone between the spring and the outlet. In fact, during floods, rain water rapidly infiltrated (new waters) whereas runoff remained very low because of the dense and permanent vegetation cover. The rapid infiltration caused a "piston flow" of the pre-event groundwater near the surface. This water is pushed out again ("exfiltration") at the base of the slope or at the swamp margin, at the level of a "dripping surface" presenting a rather low vertical fluctuation between the dry and rainy season (Gilham, 1984; Ladouche, 1997; Ndam Ngoupayou, 1997; Idir *et al.*, 1999; Ladouche *et al.*, 2001; Braun *et al.*, 2002). This phenomenon is accompanied by the release of elements such as Al, Fe, Ca, Mg, K, DOC and SiO_2. Organic material is greatly represented in the swampy zone in the form of DOC. This organic material has a very important effect on the solubilization and the trapping of the mineral elements mentioned above, in periods of floods and low water (Oliva *et al.*, 1999; Braun *et al.*, 2002). Hence, this explains the high concentration in these species in the swampy zone of the low grounds. The acidity of water which flows at the outlet of the basin is equally controlled by its organic material content. This explains the opposing behaviours of pH and DOC in the course of the flood. The very high value of dissolved silica at the low grounds could be due to a biogenic origin (silica stocked in the form of phytolytes during the degradation of palm trunks and leaves, well represented at the low grounds (Olivie-Lauquet, 2000; Braun *et al.*, 2002)). This zone of the swampy low grounds thus appears as the engine of the hydrological and hydrochemical functioning of the Mengong SEW, consequently playing a very important role in the phenomena of remobilization (stocking / unstocking) of mineral and organic elements.

CONCLUSIONS

At the Mengong SEW, the contribution of waters from different reservoirs with distinct characteristics (hilly zones and swampy zones) renders the determination of the origin of flows of water and material at the outlet of the catchment in periods of flood very complex. Nevertheless this note enables us to bring several answers to this difficulty, thanks to the association of two classical methods of study, one of which is hydrological and the other, hydrochemical.

In the hydrological approach, the quasi instantaneous response of the watershed to precipitation, the absence of the evolution of discharge at the spring and the presence of a dense vegetation, which prevents immediate runoff, enables the existence of important contributing zones to be put into evidence (swampy and exfiltration zones) which are situated at the proximity of the stream. Their contribution during floods represents >96% of water flow.

The hydrochemical approach shows that the chemical composition of water which flows at the outlet of the catchment during floods, is very close to that measured in the surface groundwaters of the swampy zone. This confirms the results of the hydrological approach. The principal mechanism of runoff generation is the translatory flow or "piston flow" under the influence of new precipitated waters. These zones are also the important areas for the processes of stocking and unstocking of chemical elements (Ca^{2+}, Mg^{2+}, SiO_2, K^+, DOC) resulting from the weathering of rocks and soils and decomposition of organic matter.

REFERENCES

Ambroise, B. (1988) Interactions eaux souterraines—eaux de surface dans le bassin du Ringelbach à Soultzeren (Hautes–Vosges, France): rôle hydrologique des surfaces saturées. In: *Interaction Between Groundwater and Surface Water* (ed. by P. Dahlblom & G. Lindh) (Proc. Int. Symp. IARH, Ystad (S), 30/05–03/06 1988, Univ. Lund (S)), 231–238.

Braun, J. J., Dupré, B., Viers, J., Ndam-Ngoupayou, J. R., Bedimo-Bedimo, J. P., Sigha-Nkamdjou, L., Freydier, R., Robain, H., Nyeck, B., Bodin, J., Oliva, P., Boeglin, J. L., Stemmler, S. & Berthelin, J. (2002) Biogeohydrodynamic in the forested humid tropical environment: the case study of the Nsimi experimental watershed (South Cameroon). *Bull. Soc. Géol. France* **173**(4), 347–357.

Dunne, T. & Black, R. D. (1970) Partial areas contributions to storm runoff in a small new England watershed. *Water Resour. Res.* **6**(5), 1296–1311.

Etchanchu, D. & Probst, J. -L. (1986) Erosion et transport des matières en suspension dans un bassin versant en région agricole. Méthode de mesure du ruissellement superficiel, de sa charge et des deux composantes du transport solide dans un cours d'eau. *C. R. Acad. Sci. Paris*, 302, série II, 1063–1068.

Gascuel-Odoux, C., Mérot, P ., Crave, A., Gineste, P., Taha, A. & Zhang, Z. (1998) Les zones contributives de fonds de vallée : localisation, structure et fonctionnement hydrodynamique. In: *Agriculture intensive et qualité des eaux* (ed. by C. Cheverry), 109–129. Science Uptade, INRA, France.

Gilham, R. W. (1984). The capillary frange and its effect on water-table response. *J. Hydrol.* **67**, 307-324.

Idir, S., Probst, A., Viville, D., Probst, J. L. (1999) Contribution des surfaces saturées et des versants aux flux d'eau et d'éléments exportés en période de crue : traçage à l'aide du carbone organique dissous et de la silice. Cas du petit bassin versant du Strengbach (Vosges, France). *C.R. Acad. Sci., Paris, Sc. Terre et planètes* **328**, 89–96.

Ladouche, B. (1997) Etude des flux hydriques par le traçage isotopique naturel à l'échelle d'un bassin forestier (Strengbach, Vosges). PhD Thesis, University of Paris VI, France.

Ladouche, B., Probst, A., Viville, D., Idir, S., Baqué, D., Loubet, M., Probst, J. -L. & Bariac, T. (2001) Hydrograph separation using isotopic, chemical and hydrological approaches (Strengbach catchment, France). *J. Hydrol.* **242**, 255–274

McDonnell, J. -J., Bonell, M., Stewart, M. K. & Pearce, A. J. (1990) Deuterium variation in storm rainfall: implication for stream hydrograph separation. *Water Resour. Res.* **26**(3), 455–458.

Mérot, P. (1988) Les zones sources à surface variable et la question de leur localisation. *Hydrol. Continent.* **3**(2), 105–115.

Ndam Ngoupayou, J. R. (1997) Bilans hydrogéochimiques sous forêt tropicale humide en Afrique: du bassin expérimental de Nsimi-Zoétélé aux réseaux hydrographiques du Nyong et de la Sanaga (Sud-Cameroun). PhD Thesis, University of Paris VI, France.

Ndam Ngoupayou, J. R., Braun, J. J., Boeglin, J. L., Meybeck, M., Sigha Nkamdjou, L., Sighomnou, D. & Bédimo, J. P. (2002) Caractéristiques hydrologiques et géochimiques d'un bassin élémentaire en zone tropicale humide d'Afrique: le bassin versant élémentaire de Nsimi-Zoétélé (Sud-Cameroun). In: *FRIEND 2002—Regional Hydrology: Bridging the Gap between Research and Practice* (ed. by H. A. J. van Lanen & S. Demuth) (Proc. fourth FRIEND conf., Cape Town, March 2002), 381–388. IAHS Publ. 274. IAHS Press, Wallingford, UK.

Olivie-Lauquet, G., Allard, T., Bertaux, J. & Muller, J. P. (2000) Crystal chemistry of suspended matter in a tropical hydrosystem, Nyong basin (Cameroon, Africa). *Chem. Geol.* **170**, 113–131.

Oliva, P., Viers, J., Dupré, B., Fortuné, J. P., Martin, F., Braun, J. J. & Robain, H. (1999) The effect of organic matter on chemical weathering : study of a small tropical watershed, Nsimi Zoetele, Cameroon. *Geoch. Cosmochim. Acta* **63**, 4013–4035.

Pionke, H. B., Gburek, W. J. & Folmar, G. J. (1993) Quantifying storm flow components in a Pennsylvania watershed when ^{18}O input and storm conditions vary. *J. Hydrol.* **148**, 169–187.

Probst, J. -L. (1992) Géochimie et Hydrologie de l'érosion continentale. Mécanismes, bilan global et actuel et fluctuation au cours des 500 derniers millions d'années. *Sci. Géol. Mém.* **94** Strasbourg, 167 P.

Toteu, S. M., Van Schmus, W. R., Penaye, J. & Nyobè, J. B. (1994) U-Pd and Sm-Nd evidence for Eburnian and Pan-African high-grade metamorphism in craton rocks of southern Cameroon. *Precamb. Res.* **67**, 321–347.

Viers, J., Dupré, B., Polvé, M., Schott, J., Dandurand, J. P. & Braun, J. J. (1997) Chemical weathering in the drainage basin of a tropical watershed (Nsimi-Zoetele site, Cameroon): comparison between organic-poor and organic-rich waters. *Chem. Geol.* **140**, 181–206.

Key word index